On God, Space & Time

On God, Space & Time

Akiva Jaap Vroman

Routledge
Taylor & Francis Group

LONDON AND NEW YORK

First published 1999 by Transaction Publishers

Published 2017 by Routledge
2 Park Square, Milton Park, Abingdon, Oxon OX14 4RN
711 Third Avenue, New York, NY 10017

First issued in paperback 2018

Routledge is an imprint of the Taylor and Francis Group, an informa business

Library of Congress Catalog Number: 98-40493

Library of Congress Cataloging-in-Publication Data

Vroman, A. J. (Akiva J.)
    On God, space, and time / Akiva Jaap Vroman.
        p. cm.
    Includes bibliographical references and index.
    ISBN 1-56000-397-9 (alk. paper)
    1. Theism. 2. Space and time. 3. God (Judaism)—History of doctrines.
I. Title.
BD555.V76   1998
181'.3—dc21   98-40493
                                                                        CIP

ISBN 13: 978-1-138-51271-9 (pbk)
ISBN 13: 978-1-56000-397-7 (hbk)

Contents

Acknowledgments

Shortly before his death, on October 24, 1989, my father asked my husband Benjamin Levy, an artist, and me to ensure that his manuscript would be published after he was gone. His request was a great honor for both of us. A few years earlier, Benjamin, who had a unique relationship with my father, developed an interest in my father's writings and philosophical theories which my father usually kept to himself. As a result, we began having discussions at home with my father in which I became very involved.

Soon after, my husband and I organized discussion groups in New York City and Tel-Aviv where my father presented his scientific theories about God to various individuals in the fields of philosophy and theology. The last time my father spoke to the editor he worked with, Jeffrey Shapiro, his words were, "Our work will be completed on October 26th." He did not realize how prophetic his words would be for that was the day of my father's funeral.

Eventually, Benjamin and I took the manuscript to the United States. *On God, Space and Time* represents about a third of the material contained in the manuscript. When we started working with Transaction, through our dear friend Professor Haim Shaked, we began gathering all the necessary information with the help of my mother, Gonny Vroman, and my sister, Yemima Vroman Ergas, who contacted some of our father's former colleagues asking for their input.

I would like to take this opportunity to thank everyone who helped us keep our promise to my father: Professor Yaacov K. Bentor, for his close partnership with my father and contribution to the Introduction; Dr. Ofer Levy, my son who is so much like my father, for his editorial and scientific insights, who began drafting the Introduction with me between working at the hospital and attending to Sharon, his wife, who was in labor at the time; Alex Grey, the editor, for his meticulous work; and Amalia G. Pena, a dear family friend, for helping me with redrafts of the Introduction.

Hanna Vroman Levy
New York, October, 1998

Introduction

Akiva Jaap Vroman was born in 1912 in Gouda, the Netherlands.* The highly gifted son of two secular high school teachers, one in mathematics, the other in physics, the young Akiva was pressed to excel academically. As a child, his parents took him on long hikes in the Swiss Alps where already the young Akiva started showing an interest in rocks. He studied geology, as well as theology, at Utrecht University, and participated in the University's theater groups. At this time, making a break with his family, he developed an abiding commitment to Zionism, and joined the Zionist movement. *On God, Space and Time*, an integration of science, philosophy, and theology, represents the culmination of Professor Vroman's work and thought.

In 1936, Akiva, having resolved to carry out his work in the Holy Land, went to Palestine, basing his geological work at Zichron Yaakov. Collecting rocks with a team of Jewish and Arab assistants, he outlined the geological history of the Carmel Mountains. At this time, Akiva became acquainted with Professor Leo Picard, head of the Department of Geology at Hebrew University, at Mt. Scopus, Jerusalem. Subsequently, Akiva returned to Utrecht University to complete his doctorate and, in 1939, published his thesis, entitled "Geology of the Region of Southwest Carmel (Palestine)." He became engaged to Gonny Betsy DeLeo who was preparing to become a Montessori teacher.

In 1940, with war erupting in Europe and Hitler's army pressing into the Netherlands, Dr. Picard invited Akiva, now Dr. Vroman, to work as a geologist in Palestine. The letter extending the invitation ensured that Dr. Vroman would receive an "entrance-certificate" into Palestine, which was then under British rule. Unfortunately, the letter was addressed by Dr. Picard's secretary simply to "Gouda," with no country name, and traveled the globe for four months before reaching its intended destina- ·

---

*The information set forth here is based, in large part, on *MONOgeoGRAPHY*, the personal memoirs and history of the geological mapping of Israel written by Akiva Jaap Vroman and published by Haifa University 1984. The authors of the introduction are the daughter and grandson of Professor Vroman.

tion, which delay very nearly cost the young couple their lives. When they learned they could both enter Palestine on the same "certificate" if they were wed, the young couple married and left for Palestine within days of receiving the letter.

Once there, Dr. Vroman joined Dr. Picard at Hebrew University where he continued his work in analyzing rocks, including rocks from all parts of the Middle East at the request of the British who, due to the war, were unable to send the rocks abroad. Dr. Vroman's three daughters, Hanna, Yemima, and Ariela were born in Jerusalem.

Between 1945 and 1948, often walking for long distances in the desert with less than the basic necessities, Dr. Vroman searched for oil in and around Ein Gedi on behalf of the Jordan Exploration Company. As the company's field geologist, Dr. Vroman created maps of the region, including Massada and Sodom.

The Vroman family lived out the War of Indeppendence in Jerusalem. Durin the war, Dr. Vroman, worked for the Israeli Army's Mapping and Photography Services.

On the day Jerusalem was freed from siege, Dr. Vroman was asked to join Dr. Yaacov K. Bentor on a secret mission for the Israeli Army Science Corps, the "Negev Oil & Mineral Intelligence," conducted at the Weitzman Institute. The mission, which was conducted over a period of ten years, entailed exploring the Negev desert in search of oil, springs, and mineral deposits. Dr. Vroman saw it necessary to redraw the maps of the whole Negev because the maps of the region made by the British were prepared using antiquated methods of mapping. Again, Dr. Vroman spent many months under difficult conditions of the desert and war.

Dr. Vroman, an excellent draftsman, drew his maps by hand, as he had throughout his entire career, by using different colored pencils to indicate the various minerals and layers of rocks. Dr. Bentor prepared the accompanying text explaining the maps and the results of their research.

In 1955, Dr. Vroman won the Israel Prize, the highest honor the country could bestow, in science for mapping the Negev. Dr. Vroman shared the prize with Dr. Bentor, who has said of him: "Akiva Vroman was a very modest person, but everyone who knew him was impressed by the breadth of his interests. He also was endowed with an unusual sense of humor." Gonny Vroman, herself an accomplished pianist who often played classical music in the adjacent room while he worked, always found her husband's ability to concentrate so extraordinary that it never allowed the rest of the world to distract him from the task at hand.

Dr. Vroman also participated in the building of the road from Beersheba to Eilat. Beginning in 1950, when the Vroman family moved to Haifa, he taught at the Technion, Israel's leading engineering college. In the early 1950's, Dr. Vroman participated in the important endeavor of bringing water from the Jordan River and rain water to a reservoir in the Beit Netufa Valley to irrigate the Negev.

In 1953 and 1954, Dr. Vroman was asked by Lapidot, an Israeli oil company, to compile detailed maps of all of Israel and identify promising areas for drilling oil. A result of this work was the discovery of oil and gas in Hulikat and Rosh Zohar. Dr. Vroman made an invaluable, central contribution to Israel's search for oil through his pioneering development and use of aerial photographic techniques which, beginning in 1947, he had taught himself. These techniques enabled him to map efficiently and accurately the areas of the Negev which had previously been poorly mapped by the British.

In 1956, Dr. Vroman went to Mexico as the delegate of the Geological Institute to the 20th International Geological Conference. After the conference, on a fellowship awarded him by UNESCO, Dr. Vroman traveled to Harvard University to further his knowledge of geological structures spending most of his time in libraries copying geological maps of Arab countries.

Using the aerial photographic techniques he had developed during the Sinai Campaign, in 1957, Dr. Vroman prepared a geological map of the Sinai demonstrating how maps of otherwise inaccessible areas can be drawn efficiently and accurately. After the Six Day War, the maps created by Dr. Vroman, a pioneer in Israeli geology, were used by the Israeli Army which found them more accurate than those previously made by British and Egyptian geologists who had prepared theirs by working on foot.

Many years later, his friend and colleague Y. Bartov would say of him: "It was Vroman who recognized the vast possibilities of geological mapping by means of aerial photography. He did so at a time when aerial photography was almost unknown in the country, in any case, not for mapping purposes, and when in world-wide terms the branch of photo geology was still in its infancy.... With the barrier [of antiquated mapping methods] thus removed, there was no turning back."

As head of the Israeli Geological Mapping Department, in 1958, Dr. Vroman, now a professor, set about to create maps of Mt. Carmel and the region of the Galilee.

Professor Vroman's work extended to Africa and throughout the

Middle East resulting in maps as well as a published book on the tectonics of these areas which offered many original ideas. He once described these maps as the "instrument" of the scientific conclusions he set forth in *A Reappraisal of the Structure of the Earth*, which was published in 1973.

In 1964, as the first Israeli to do so, Professor Vroman went to France to study additional aerial photographic techniques. During the 1960s and 1970s, he continued to teach, among other subjects, his pioneering mapping techniques to a generation of Israeli geologists at Haifa University, Tel-Aviv University, and Beit Berl. During this period, he also published several scholarly articles and books on a broad range of subjects. His regional maps and his compiled map of Israel are still current and used by geologists today.

Professor Vroman retired in 1979, and devoted the last ten years of his life to philosophy. He started to write about his scientific concept of God. During this period of reflection and writing, Professor Vroman, a true Renaissance man, also found time to pursue his passions in music and art, creating some very beautiful drawings. As happens to many scientists in their mature years, he began questioning the creation of the universe. His long days of mapping connected him physically, intellectually, and spiritually with the terrain of the Holy Land and the region. Expanding the scope of his scientific work from the Holy Land to include the Sinai, Arab countries and Africa had led him to a "reappraisal of the structure of the earth." This journey took him from examining rocks under a microscope to developing a scientific theory of how the universe came about.

Professor Vroman had very high moral standards which he preached and lived by. These standards were based upon what he believed was the most important message in the Bible and *On God, Space, and Time*: "Love thy neighbor as thyself."

Hanna Vroman Levy and Ofer Levy

1

An Introduction to Reality and Imagination

We say a thing is real when we are convinced that it exists independently of the idea we have formed about it.

The term "exists" has been a source of much misunderstanding. To state that a thing "exists" does not contribute anything new about its characteristics. The German philosopher and mathematician, Gottlob Frege (1848–1925), counted existence among "second order predicates." The statement "horses exist" belongs to the category "horses are numerous," and it does not tell us anything new about horses per se. There have even been philosophers who included existence with occupation, and made statements such as "existing things are engaged in existing." But the nonsensical statement "unicorns do not exist, they have better things to do" may open their eyes.

The term "existence" has very often been confused with "essence." Essence delimits an object by its definitions; but even an exhaustive description of all its properties does not imply that the object really exists. We may imagine hobgoblins, but they will not be real—even if we ascribe to them all the attributes that may cross our minds, down to the last invented detail. The essence of hobgoblins has therefore nothing to do with their existence. But it is hardly possible to prove that hobgoblins do not exist. Tomorrow we may be shocked when we meet a real hobgoblin!

But this is an old story: the reasoning of German philosopher Immanuel Kant (1724–1804). He wrote: "Whatever and however much our concept of an object contains, we must go beyond it in order to ascribe existence to it."[1] Nevertheless, we may demonstrate that our mind may imagine things that do not exist, and that we may even provide proof that they do not exist.

We are able to give free play to our imagination. We may keep our minds on a star of a galaxy in the remotest part of the cosmos, and entertain the illusion that it is over there in exactly the same state as we

1

now physically see it. However, we must be well aware that this is nonsense because of the finite and constant velocity of light.

But we can just as well imagine a day in an infinite past as being an infinite number of days behind us—such as our day of birth. Our next step is to inquire if such a day in the infinite past can really exist. For example: we can check the next day and all the following days that move towards the present. The day in our minds becomes tomorrow, and after today it will retreat into the past and become yesterday. Then after the lapse of another day it would become the day before yesterday.

Time passes and we would expect that the lapse of every additional day would bring our day closer and closer to the day of our birth, to the point where the two days will coincide. But this first day of our existence will never arrive, for we assumed at the very outset of our reasoning that an infinite number of days separates the day we kept in mind from the day of our birth.

You must keep in mind that the passage of time does not skip even one day. Time counts all the days. There would be no end in time's counting of the succession of days if the counting would start at a day in an infinite past. The days would lapse, but the dawn of known history would never rise, and we and all previous generations would never have been born.

A day in the infinite past is therefore nonsense. So we are forced to accept the only alternative: all the days that have elapsed belong to a past of countable days. They belong to a past that started on a first day a finite number of days ago. Time was created with this first day. And, as we do not call our own existence into question, it follows indeed that an eternal past does not exist. Time must have a beginning. This reasoning is based on a simple mathematical law that is often ignored.

An infinite quantity remains the same infinite quantity if a finite quantity—however large—is subtracted from it.

Let us apply this law to our day in the infinite past. An infinite number of days has to elapse before the day of our birth will break. This infinite time will never decrease, hence it will never become finite and then become zero. This follows from the simple reasoning that the duration, which is the total length of elapsed time measured from our chosen starting point, can never grow into an actual infinite eternity. It must always remain a potential infinite duration.[2]

Time is not an independent parameter. We cannot imagine velocity without time. Velocity is the time in which a moving object covers a certain distance. There is no movement without time, and in the absence of time all the objects would be at a standstill. Stop reading for a moment and take a minute to try and imagine a world without movement. The atoms in crystal latices would not oscillate with their clockwork precision, and would be in state of freeze at absolute zero degrees. Light would be absent, planets would not orbit around the stars, nor the stars around the neighboring stars. This would be an absurd situation. The force of gravity will draw all matter to one point and into naught. But this is equally absurd because such a process would need time which did not exist. The world is unthinkable without time. The world has a beginning, and it began to exist the moment time came into existence.

The alternative is that there has been a time-interval, a duration between the moment time came into existence and the moment when the world came into existence. This is a strange supposition because we are unable to comprehend a time-interval that in the absence of moving and changing objects is immeasurable in principle.

Let me elaborate: Our mind is a rather odd machine which knows from experience that the world is not static, but that it changes all the time. On the other hand it refuses to accept changes—it deems them arbitrary and whimsical. Just as we do not accept different things on the two sides of the equal sign of an equation, we tend to eliminate change by accepting a cause that brings about the effect. Each cause is linked to its effect by time interval. The law of cause and effect is unthinkable without time. This conflicts with the law of logic. Time does not play a part in mathematical logic. If A equals B, A does not equal C after B equals C, but because B equals C.[3]

It is clear that an equilateral triangle did not obtain three angles of sixty degrees after the sides became equal, rather the triangle is equilateral because the three angles are equal. In contrast, water boils after it has been put on the fire. But a causal link is not merely a link of temporal sequence. Our mind turns the "after" into a "because," which means that we infer that putting water on the fire brings about its boiling.

Our conviction that this causal link exists is not enhanced by repeating this experiment continuously. Professor Jean Piaget stated, correctly, that this conviction does not depend on our sense perceptions. The law of Leucippus[4]—that every change must have a cause—is something innate in our mind. It is a law of thought on an equal footing with our other laws of thought, such as the laws of thought established by Plato and Aristotle—the so-called three laws of logic. Immanuel Kant called these laws *a priori* synthetic judgment.

The great miracle is that in our daily lives we are never disappointed by experience. The course of events always complies with our anticipation—in this case the anticipation that water will boil whenever we put it on the fire.

Why indeed should the external world be concerned with the laws of our anticipation? The mental law of cause and effect is so much part and parcel of our mind that it was taken for granted until the twentieth century. It is a law that has to be reappraised with the discovery that Kant's synthetic *a priori* is a valid mental instrument under normal circumstances; but that the rules break down in the world of very high speeds—of very large gravitational forces—where the geometry of space is curved. And equally so in the world of nearly infinitesimal smallness of subatomic dimensions, where the speeding particles have no exact location and may even exchange identity when one particle crosses the fairway of another.

Notes

1. *Critique of Pure Reason.*
2. A potential infinite is defined as a finite that grows and grows but can never become an actual infinite—although the end of its growth is never in sight.
3. The science of dealing with our innate ways of thinking is called epistemology. Immanuel Kant was one of its pioneers and the modern psychological aspects were studied by the Swiss psychologist, Jean Piaget (1896–1980.) Piaget was a professor of child psychology and wrote numerous books on the subject. See Bibliography.
4. Leucippus, a Greek philosopher, lived during the fifth century B.C. It is believed he proposed that matter is made up of indivisible and infinitely small atoms. He is regarded as the mentor of Democritus (c. 460–370 B.C.); but he lived in a past so dim that some doubt he ever existed.

2

The Modern Vindication
of the Existence of the Creator

That the world must have a beginning and must have been created out of nothing, the *Creatio ex Nihilo*, was already discovered in the early Middle Ages and followed the lines of logical and causal thinking; that is, not all from the consequences of our sense perceptions (modern physical experiments did not exist then). It was the first instant that it was discovered that our innate laws of thinking are, in a way, deficient—that impeccably rational thinking leads to an irrational conclusion. That everything we perceive has been created out of nothing. The reasoning, nearly 1500 years old, opened with the book *Contra Aristotelem* by the monk John Philoponus. The issue was taken up anew by the Muslim Kindi and finally brought to perfection by Saadyah Gaon (882–942).[1]

William Lane Craig described this history in *The Kalam Cosmological Argument*. This discovery was of tremendous theological importance, because of the association of the *Creatio ex Nihilo* with the principle of a First Cause. A term is very often used as a definition for God. God is a Something we experience as Nothing that created everything out of a Something—the same Something we experience as Nothing.

It took the medieval Cosmological Argument (CA) of the Kalam School[2] three centuries to ripen. Then it sank into a long oblivion of thousands of years. This misfortune was caused by the very man who started the argument, Aristotle (384–322 B.C.). The Greek philosopher was well aware that actual infinites do not exist. He knew they only exist in the mind of the mathematician; but he believed without any practical reason that he was permitted to exempt time from this rule. He disbelieved in the real existence of time, and I have summed up his superstition with the following poem:

5

The past we forget
the future is not yet
the present is their joint
but it is just a point

Hence he believed that there is nothing real in time—it exists in the imagination.

The second factor that buried the Kalam Cosmological Argument for so many years was the great respect for Aristotle among the thinkers of the Middle Ages. The CA was regarded as a revolt against Aristotle's garbled thinking about time. And it was so indeed. The modern view is that everything exists which is measurable, and that there is no other parameter that may be measured with greater precision than time. It is, moreover, a fact of life that there is no other parameter that kills more cruelly than the lapse of time—aging.

After a thousand years the Kalam Cosmological Argument has been resurrected as an archaeological curiosity by Dr. Herbert Davidson and by Harry Austryn Wolfson. William Lane Craig immediately recognized the enormous theological consequences.[3]

Albert Einstein (1897–1955) developed his own vision of time. He believed that time did not lose its reality, but the rate of its flow became dependent on the relative velocity of objects. It was realized that the Maxwell-equations[4] of the electromagnetic field (light) imply that the velocity of light in space is invariable. As a consequence, the constant rate of flow of time went by the board, and a fourth linear dimension had to be added to the conventional length, width, and height that mark Euclidean[5] space to describe the movements of objects. This fourth additional dimension contains the parameter time. But our imaginative faculties cannot conceive such a four-dimensional cosmos.

Albert Einstein's original idea was that the cosmos is invariable as to its volume—it could neither shrink nor expand. But that was his conclusion until corrected (independently) by the Dutch astronomer Willem de Sitter and the Russian Alexander Friedmann who drew Einstein's attention to a miscalculation.

Einstein was broad-minded enough to admit mistakes, but this one had deeper philosophical consequences. It meant that Einstein's universe must expand continuously.

I have reason to believe that it was not the future of the universe that annoyed him, rather it must have been the consequence of the "flash back" (what happened at the beginning). A cosmos expanding out of nothing. This conclusion went against the grain of his worldview, which was purely an idealistic one, and the argument that God and nature are the same[6] had suddenly to be rejected. God could no more be identified with the gigantic clockwork of the revolving galaxies. He must have created it.

It seems that Einstein was wise enough to quit in 1923 from his view that the cosmos is static as to its volume. But in that same year his theory of relativity[7] (proposed in 1905) implied much more than the original Kalam CA. It established the creation of time and space in a distant past and it revealed an inkling about the circumstances. But it was still nothing but a brilliant theory. The world waited eagerly for corroborating facts.

The first confirmation came with American astronomer Edwin Hubble's[8] discovery in 1930 of the red shift of the spectral lines from the light of the distant galaxies. The shift of the spectral lines to the red side of the spectrum (the side of the longer wave length) may only be explained as a speeding away of the radiating celestial bodies—the galaxies. This conclusion has been contested by several skeptics, such as Patrick Shaw (logics lecturer at the University of Glasgow).[9]

Shaw warned against the following reasoning: "Let us assume that the universe is expanding. If it were, then the spectrum of the furthest stars should show a shift towards the red. This redshift is precisely what we find. So we can assume that the universe is expanding."[10] And Shaw added the following warning: "Invalid as it stands. We need a way of blocking alternative explanation of the shift."

Let us meet Shaw's request: the Dutchman Maarten Schmidt and New Yorker Jesse Greenstein calculated many years later (after 1930) that all the other known factors that might have been responsible for the redshifts—among them a very strong gravity field—are absolutely inadequate to account for the magnitude of the observed shift. Their paper has been regarded by all the former objectors against an expanding universe as "overwhelming evidence."

The discovery of the redshift was later complimented by other discoveries confirming an expanding cosmos. In 1965, American radio astronomer Arno Penzias (born in Germany, 1933–) and Robert W. Wilson (1936–)[11] of Bell Telephone Laboratories found a faint cosmic glow of barely three degrees above absolute zero temperature. It bears the characteristics of a so-called black body radiation—a feature that on a cosmic scale may only be explained as a late stage of the ever declining effect of an event of radiation of enormous proportions in the most distant past. It is the last sign of the "big bang," as the fireworks of the creation of the universe is popularly called...following from the model proposed by Belgian physicist Georges Edouard Lemaitre (1894–1966).

Another feature that supports the conclusion that the universe is expanding is implied in an old enigma: why is the sky black at night? The amount of starlight must be so great that the night should be as bright as the day. Heinrich Wilhelm Olbers (1758–1840)[12] asked this question—known as Olbers' Paradox—more than a hundred years ago and could not solve the problem. Today the answer is clear: the universe needs an enormous quantity of energy to expand, and it borrows this energy from the radiation-energy of the stars.

A third and most striking discovery is the character of the so-called quasars studied by Allan Sandage in the late 1980s. The most distant objects in the universe do not bear the characteristic features of the much closer—and apparently normal galaxies—which turned out to be composed of billions of normal stars. The brightness of just one quasar is enormous. It outshines the light emitted by an entire group of galaxies by a factor of ten. Their redshift is outermost; they are extremely remote. It follows that the speeding away of any object in the universe is directly proportional to the distance from that point. The greater the distance, the greater the speed, and the greater the redshift.

Furthermore, one should realize that the radio telescope, through which the quasars are observed, not only looks very deep into the universe—it also looks far back into time. The velocity of light is constant but not infinite, and the quasars are so remote (the most distant objects we have yet seen) that their light has to travel billions of years before it reaches our eyes. We observe quasars not in their present state, but as they were billions of years ago. When we look at them we feel as though we were present at the very moment of the birth of the cosmos—a hot glowing crucible out of which the galaxies emerged.

Let us perform the following mental exercises: let us imagine people standing at this very moment on a celestial body we observe as a quasar. This is not an unreasonable proposal because our looking back in time means that the object we see is at this moment billions of years older than the quasar we see through the radio telescope. The people we imagine standing on it have the entire age of the universe behind them, just as we have. They might find themselves standing on a life-supporting planet like our earth. They peer into a telescope in their quest to explore our Milky Way and point it at us and lo! they do not see us standing on earth rotating around our sun. They do not even see the Milky Way, which is our galaxy. They just see another quasar, for it is now their turn to look back in time.

Scientists rarely fix their eyes on the first cause. They restrict themselves generally to theoretical details of the first effect, which is the creation of the world—the big bang. A more sophisticated name for the big bang is the *space-time singularity*.[13] A singularity is an object of infinitesimally small diameter and of an infinite mass. These two characteristics are—as I pointed out—taboo from a philosophical point of view. Paul Davies said in the preface to *The Edge of Infinity* that a singularity is rather "a non-place where all known laws are suspended."

This taboo, which we deduced above, is the following: actual infinitesimals and actual infinites do not exist. However, Davies made it clear in chapter 4, "Towards the Edge of Infinity," that nature does not need to transgress this taboo in order to create objects featuring all the strange side-effects that theoretical physics attribute to singularities. It only depends on their mass: the greater the mass the lower the critical density. For example: a star group of a million suns (not an outlandish feature in the skies) might collapse by gravity to a modest density of one kilogram per cubic centimeter. This is quite normal among ordinary stars and the whole compact mass would behave as a singularity of infinitesimal dimensions and infinite density.

I am thus strongly tempted to modify the definition of an actual singularity as actual bodies that behave as if they are infinitesimally small and of an infinite density. If armchair theoreticians would maintain that further collapse to infinite density is unavoidable, I would counter that there must be something amiss with our knowledge of matter.

Singularities crop up in purely mathematical calculations and emerge in two varieties: invisible Schwarzschild singularities, commonly called black holes, and fiercely radiating naked singularities. A black hole is a singularity encased within a surface marking the radius, or the distance from the singularity where the emitted light turns back into the singularity because of the tremendous gravitational attraction it exerts on the emitted light beams. This surface is called the "Schwarzschild horizon."

It has been reasoned by Paul Davies that a singularity rotating at a high speed, or one charged with a very large amount of electricity, has no Schwarzschild horizon. Such a singularity is termed naked. Stephen Hawking pointed out that it follows from the laws of sub-atomic physics and from quantum mechanics that every singularity, whether attired in a Schwarzschild horizon or not, must emit some energy. The more energy emitted, the narrower the radius of the shield.

Hawking figured in 1974 that all the tiny black holes must have evaporated long ago and that hardly any leftovers of the space-time singularity—the big bang—are still alive. Further calculations lead to the conclusion that the force responsible for the creation of new black holes is gravity. Stars thirty times larger than the diameter of our sun must collapse under their own weight to an infinitesimally small point of infinite density.

The most conspicuous phenomenon of a collapsing, imploding, super-massive star is the supernova state[14]—the explosion of the outer shell of a giant star that collapses under its own weight, creating a tremendous amount of heat that causes a violent casting off of its outer sphere into space. New elements are created in the process—carbon among them, the element of our life-spring. The duration of one single supernova event is at least a billion years. The history of the supernova event is supposed to begin with a star that swells to giant hydrogen, becoming a red supergiant of 170 million degrees Celsius when gravity squeezes it, and its hydrogen atoms are fused into helium. The helium atoms are then fused to become carbon and oxygen. Nuclear fusion continues and carbon is bound into neon. Meanwhile the temperature rises to one and a half billion degrees. This occurs seven years before the last stage and oxygen is gradually turned into silicon. Four days before the great event silicon is bound into iron, the most stable among the chemical elements.

Further collapse cannot release more energy by simple fusion. The very cores of the iron atoms are crunched, releasing terrific amounts of the tiniest particles called neutrinos, which rip through the surface and cause a most violent shock wave. More than 99 percent of the star's energy is cast off.[15] What remains at the core is what is called a black hole. Observed facts suggest that giant black holes rotate the luminous gas-arms of the quasars. Our knowledge about the big bang would be greatly increased if we knew more about singularities. Much of what I explained here is pure armchair science. What counts is the verification of all these theoretical deductions by facts in the physical world observed through our senses.

About twenty supernova are found each year in the more distant galaxies. We had the good luck to witness a much closer one on 23 February 1988 in the "Large Magellanic Cloud" at a distance of about 170,000 light-years from Earth. The message arrived that the blue star Sanduleak 69 degrees 202, which had a mass twenty times that of the sun, blasted into supernova. The analysis confirmed the expectations, but the densest remnant of a supernova ever found was not a black hole or other singularity, but a body of a stage less dense called a "neutron star." Special types of neutron stars are called "pulsars." They emit an x-ray beam rotating like the light of a lighthouse. Neutron stars are composed of tightly packed building-stones of atom-cores—neutrons. Their density is so tremendous that a small marble of neutron stuff would weigh ten billion pounds. It's not a toy to play with, but nevertheless it is of a density that falls short of the infinite.

As to the cores of quasars: I mentioned above that practically all scientists are convinced that such phenomena are super-massive black holes spinning like a turbine and gobbling up spiraling strands of gases that lit up a billion years after the big bang. But even the strongest conviction cannot compete with solid observed facts, and they are few. Quasars are too old and too far away to study in detail. Is there not a much closer and more suitable heavenly body such as the core of a nearby galaxy that may reveal the same features as the cores of quasars, though on a much more modest scale. Yes, the nearest one is the core of our own galaxy, the Milky Way, only 15,000 light-years away. Only the latest piece of equipment, the "Very Large Array" radio telescope near Soccoro in southern New Mexico is able to peer into the overcrowded, glowing center of the Milky Way.

Scientists agree on the facts, but are not absolutely unanimous as to their interpretation: A giant central rotating gaseous shell of some ten to six light-years radius. It gobbles up spiraling streamers from a molecular cloud and whips them around at a velocity of ten kilometers per second on the periphery; and accelerating to 200 kilometers per second close to the central shell. If there was no giant black hole at the very center, the spiraling gas filaments would have flown off their paths long ago. Common sense tells us that the core of the Milky Way is a textbook example of a mini-quasar.[16] It is the closest encounter with a true black hole.

But still there are other scientists who disagree. Such as the group around Hannes Alfven. Here follows a short account of their views:

> One calls the rotating ionized (electrically charged) molecular gases, such as around the cores of quasars and galaxies, a 'plasma.' This plasma may continuously be in a state of being discharged by an electric arc. This bolt produces a magnetic field at a right angle to the current. Laboratory experiments—computer models, and actual astronomical observations of the core of the Milky Way—confirm that the magnetic field causes the straight current to wobble and to turn it into a spiral.

Eric J. Lerner, a freelance researcher in Lawrenceville, New Jersey, deduced from his experiments that the first small random currents cause the plasma to curdle into clods and that the clods tend to rotate around the merging currents. They are then drawn out into spiraled filaments, forming a pattern that brings spiraled galaxies to mind.

I am always of the opinion that one should disregard the notion that dissidents are always in the minority. What counts is their arguments and how to deal with them, even when the proposals may turn out to be the mere hallucinations of some cranks. I even listened with attention to an old man, a certain E. Wexler, who defended the untenable position of the French scientist, mathematician, and philosopher René Descartes (1596–1650)[17]—that the universe is ruled by the churning of violent ether-storms. But Hannes Alfven and some of his colleagues are certainly not cranks. They are lucid of mind and serious scientists who propose electromagnetic phenomena as an alternative to the force of gravity. On the other hand, the force of gravity rules the history of the universe according to the current standard model of the big bang. The most outrageous teaching of their model of electromagnetic plasma is that the universe should have an eternal past. We have seen that simple reasoning rules this absurdity out, though many agree that electromagnetic forces may have had some influence on the history of the cosmos. There is much evidence that electromagnetism is too weak to push matter around on a cosmic scale. I propose, therefore, that the plasma theory may explain the early germination of the quasars when the universe was still very young and in a dense plasmatic state.

But neither gravity nor electromagneticism can explain the enormous velocities reached by the fringes of the spiral nebulae's arms turning at a speed which seems to defy Newton's simple relations of the law of gravity. There is, therefore, still a frantic search in the skies for additional heavy matter that should account for this anomaly.

After the discovery thousands of years ago that time must have been created, and the hardly challengeable new discovery of the big bang, we have become too smug. We even disregard the most basic truth that the smartest intellect is unable to decipher the objective reality from the poor information we receive from our few and deficient senses.

All these discoveries, theoretical and practical, clinch the CA. Many theologians in search for God would certainly content themselves with the evidence of a First Cause as the appropriate, and even conventional, definition of God. There are, however, many others who worship their First Cause not so much as the Creator but as the "Almighty," the "All-Good," the "Bountiful," or the "Omniscient" to mention only a few of the endless string of divine perfections mentioned in the Bible and in daily prayers.

God did not only create what we call the cosmos of time, space, and matter but the mental world as well. He is the Creator of everything, including all the perfections that adorn Him in the text of Holy Writ. Did God create His own perfections? God defined as the First Cause can have no other predicates. My aunt happens to wear a jumper she knitted with her own hands. It is her legal property, but not her predicate, not her attribute.

There are numerous scholars who are well versed in theological problems and who try to seek confirmation of God's existence through other arguments. The Ontological Argument (OA) and the Argument from Design (AFD) are two such approaches with long histories.

Notes

1. Saadyah Gaon was one of the greatest of Jewish religious authorities, philosophers, and writers. He was born in Fayyum (south of present-day Cairo) and was the first translator of the Bible into Arabic.
2. Old Jewish and Arab philosophy.
3. See Bibliography on the works of these authors.
4. Scottish physicist, James Clerk Maxwell (1831–1879), unified in theory all the phenomena of electricity and magnetism.
5. Euclidean means "of Euclid" who was born from about 330 B.C. to 275 B.C. He was one of the great Greek mathematicians, and is best known for his book, *Elements*, concerning mathematical knowledge.
6. The philosophy of Baruch de Spinoza (1632–1677) is illustrated in his expression: *Deus sive Natura*—God alias Nature. It meant that God conjured Himself out of nothing.
7. The theory of the universe, based on the principle that measures of motion, space, and time are relative (*Oxford English Dictionary*).
8. Hubble (1889–1953) discovered millions of galaxies other than our own. Hubble's Law: the velocities of receding galaxies are directly proportional to their distance from the solar system.
9. Patrick Shaw, *Logic and its Limits*.
10. Ibid. p. 189 statement 15.2.
11. Penzias and Wilson shared the 1978 Nobel Prize for physics with Russian physicist Pyotr Kapitza.
12. Olbers was a German astronomer who discovered the asteroids Pallas and Vesta and rediscovered Ceres. He also found five comets and one is named after him.
13. Paul Davies, *The Edge of Infinity*; Stephen W. Hawking, *A Brief History of Time*.
14. Rick Gore, "The Once and Future Universe," *National Geographic*, June 1983.
15. P. Kirshner, "Supernova, Death of a Star," *National Geographic*, May 1988.
16. Marcia Bartuseak, "Coming Home," *Discover*, September 1988, and "Feeding the Hole," *Discover*, June 1989. I. Peterson, "Carrying Fuel into the Galaxy Center," *Science News*, 14 January 1989.
17. Descartes, educated by the Jesuits, experimented with mathematical conclusions that he tried to make the basis of his philosophy.

3

God: The Ontological Argument and the Argument from Design

William Lane Craig concluded his work on the CA with the words: "—but whether the Creator is omniscient, good, perfect, and so forth, we shall not inquire." Indeed, I say we'd better not!

I could not find much sense either in the concluding remarks of the chapter on the Cosmological Argument in Brian Davies's (not to be confused with Paul Davies) work, *An Introduction to the Philosophy of Religion*. Davies said: "As an argument for a first cause of all existing things, the cosmological argument seems a reasonable one. But it does not by itself establish the existence of God with the properties sometimes ascribed to Him." Brian Davies failed to notice that "the properties sometimes ascribed to Him" are part and parcel of all existing things of which He is supposed to be the First Cause. Davies obviously does not understand that one cannot have both God as the First Cause and God endowed with all the divine attributes ascribed to Him because this would imply the absurdity that God is the First Cause of His own attributes.

One should not try to make the circle square. On the one hand rejoicing at the vindication of the CA, and on the other hand nevertheless continuing to worship all the divine excellences hallowed in the Bible. That those aware of this incompatibility do not change their habit is the consequence of our human foible...our irrepressible urge to project our human virtues on God.

The Ontological Argument[1] and the Argument from Denmark are often adduced to prove God's existence either independently or in conjunction with the CA, and they are not free from at least a tinge of the same incompatibility I described above.

The OA starts with a statement of God's essence and the AFD supposes that God adapts means to an end. This implies that both arguments assume that God is endowed with positive attributes. The OA was used by Archbishop of Canterbury, St. Anselm (1033–1109), to prove God's existence. He opened his theory with a statement[2] about God's essence and described Him as "something nothing greater than which can be thought."

The next sentence drove him straight into the eddies of total confusion. He dared to maintain that this statement about God's essence implies His existence. St. Anselm's famous paralogism runs as follows: "—For suppose that it (the something nothing greater than which can be thought, which existed in Anselm's fertile mind) exists in the understanding alone; then it can be conceived to exist in reality, which is greater."

I am at a loss how to attach "greatness," or any other measure, to reality or to the imagination. And I am even more confused—it is even beyond me—how to compare the measure of greatness of the imagination with the greatness of reality. This absurdity is expressed in Immanuel Kant's critical remark: "A hundred thalers in my mind is not one thaler less than the hundred thalers on the table."

As illustrated in the first chapter, I cannot conjure my imaginary green hobby horse into existence by appealing to the perfection of its greenness. I cannot claim that it would be less green if it did not exist.

Though it is perfectly clear that the saint's OA has no validity as an independent proof of God's existence, we may nonetheless attempt to save his formulation by using it in conjunction with the Cosmological Argument. The statement that the Creator is "something nothing greater than which can be thought" sounds very convincing, but as soon as we realize that this "Something" is "Nothing," we discover that ontology is a mine field.

We are aware that as mortals we are not able to translate this "Nothing" into "Something" because we cannot peer around God's corner. Our inner urge to identify our Creator with Anselm's "something et cetera" has no reasonable base whatsoever. It is more like an emotional exclamation not unlike King Solomon's pious ejaculation: "But will God indeed dwell on the earth? Behold the heaven and heaven of heavens cannot contain thee."[3]

The AFD, also called the Teleological Argument, appeals to God's intellect, which seems to pervade even the most minor details of the cosmos. To believe in God's intellect does not necessarily imply that one should believe in God as a Creator. One might even contend that God's intellect has to be identified with designed Nature.

There were a number of thinkers who were less impressed by the beauty of this design. The English philosopher and mathematician Bertrand Russell (1872–1970) wrote: "If rabbits were theologians they might think the exquisite adaptation of weasels to the killing of rabbits hardly a matter of thankfulness." A contemptuous gibe is not a valid argument in the defense of atheism (Russell's atheism); but it certainly underlines my feelings, expressed in the preface, that it appears as if the Creator is impervious to the fate of His Creation and He wished us to be our own guardians against evil.

The words of Paul Davies on the AFD are much more inspiring. He said: "When the physical conditions in the big bang are examined, it appears that none of the presently observed cosmic organizations existed at the beginning."[4] And he explains that the cosmos came into being as an infinitesimal pinpoint. A 10^{-43} second later it had a temperature of 10^{32} degrees. But this tiny, hot droplet, this speckle of lawless chaotic subatomic soup, from which the universe proceeded, was already endowed with the potential to develop all the laws of nature that characterize the present gigantic cosmos.

We cannot attribute any human intellect to the Creating Nothing. And, on the contrary, the Nothing created this intellect. I therefore propose to dispense with the term "intellect" and not to attribute to God designing activity. Let us call it a divine providence—though not of the unknowable Creator. Let us use this term to indicate that we believe to see, with our human eyes, a "design" in His work, His creation.

There is another modern version of the AFD. It is called the Anthropic Principle—a term coined by Brandon Carter in 1974.[5]

According to the Anthropic Principle, if the universe would be only slightly different, mankind would not be here. We exist thanks to an unknown number of highly improbable events that seem to flout all the laws of nature. The two big riddles are:

1. Why is the universe isotropic? (This means that its characteristics are the same in whatever direction we look.)
2. Why is the universe flat? (This means that it will not expand forever. We do not live in an open universe; nor will the universe ever collapse again into nothingness under the force of gravity. We do not live in a "closed" universe.)

The wisecrack that it has been calculated that galaxies would not have been formed and that we would not exist if the big bang had created a chaotic universe, and if the expansion did not come to a standstill, is still not an answer. Rather, it begs the question. It is indeed an unresolved riddle and very suggestive that a guiding hand has designed it that way. The huge problem illustrates furthermore that we cannot think intelligently about the universe and its history without taking the probability of our coming into existence into account.

Take an even closer look at the details: there is no doubt that the development of the human consciousness, from the inanimate states through the lower animate states, must pass though a succession of well-defined stages. We might even estimate the age of the universe if the duration of all these stages were known. But there is not the slightest chance that we would ever succeed to make even a rough guess of the duration of the stages in the development of the mind.

Astrophysicists do not concern themselves with the mind-body problem. Their study-object is the physical world and nothing else. But their findings on the development of the physical universe must have a bearing on the development of the mental reality because of this interrelation between the mental world and its physical reflection.

Astrochemists know that the big bang created hydrogen and helium in the nick of time, but not the heavier elements of the "periodic system of Mendelye'ev," and most certainly not carbon, which has a nucleus of six protons and six neutrons. Carbon is nevertheless the vital component of organic molecules, and therefore of our body. Carbon may only materialize in "supernovae." Robert Dicke (who once interpreted the origin of the "cosmic radiation background" discovered by Wilson and Penzias) commented: "It would be impossible to observe a universe younger than the shortest-lived star, because the very elements they are composed of wouldn't exist."

We know that the supernova stage of a giant star, which turned it into our sun and solar system, occurred about five billion years ago, which means that the age of the universe must be considerably older. But after this feat—by no means a very common event—a long series of other events followed, creating at the end, mankind and human consciousness. The realization of every event is so improbable from a statistical point of view that we cannot suppress the feeling that a guiding hand, which we call Divine Providence, cared for the emergence of the human mind.

We have to handle this form of the AD with prudence. If the physical universe is in reality a mental universe, it must follow that the stages leading to the creation of mankind follows a line in the mental reality, and its details must remain an eternal secret for us because we have no other tools at our disposal, but our senses, to gain information from the external world. These tools cannot supply us with anything else but a world of space and matter. And this may suggest that behind all the processes and events—which look so extremely improbable when seen under the physical aspect—very trivial mental processes may be so concealed from our senses that we will never be able to discover them.

This modest or weak Anthropic Principle is a product of sound reasoning, though it is somewhat rash to regard it as a valid AFD. But Princeton University physicist, John A. Wheeler, distorted this principle by stretching its point beyond recognition. His argument produced what is called the Strong Anthropic Principle and the Participatory Anthropic Principle. The sophism runs more or less as follows: "Reality is a creation of the mind, and the faculty of observation is thus tantamount to the faculty of creation. Man participates in the existence of the universe by looking at it."[6] The world is brought into existence by the collective observations of all observers, past, present, and future.

But it is a mystery how Wheeler, a man of good understanding and well seasoned in modern science, may become the victim of such insane fancy. It is even worse than the ideas of Irish philosopher and Anglican Bishop of Cloyne, George Berkeley (1685–1753), who dared to maintain that "existence depends upon its being observed."

These two philosophers have one negative point in common. They ignore that the source of sense perceptions is the external world stimulating our senses and that this obvious fact is adequate evidence that sense perceptions are obtruded upon us beyond our will and beyond our control—and that, therefore, the external world does not depend on the use of our eyes. It is tripe to maintain that the laws of nature exist because we behold them. Such a claim is symptomatic of the megalomania that "man is the creator the world," and not just "the measure of all things."

The only man-made elements in Creation are the theories of how the world had been created; and though Mankind's thoughts are mental reality, the contents of his thoughts are not. This is a truth that again returns us to chapter 1: we may entertain the craziest imaginations about the creation of the world, but we do not, by our fantasy, create a crazy world.

Notes

1. Ontology is the science that deals with existence and essence. Its exploration is a tricky occupation.
2. *"Aliquid quo nihil maius cogitare possit."*
3. I Kings 8:27.

4. Paul Davies, *The Edge of Infinity.*
5. Tony Rothman, "A 'What you see is what you beget' Theory," *Discover*, vol. 8,5 (May 1987).
6. The words of Tony Rothman, *Discover* magazine.

4

God, Mind, and Body: Part 1

We will show in the next two chapters that the world of space and matter is nothing but the interpretation from our sense perceptions by our working mind, and that the reality behind this interpretation is a mental external world. The arguments used as evidence that our "world of space and matter," that is, "the physical world," is indeed a reflection through our senses of a mental reality, will be elucidated in the following pages. I appeal to your willingness to accept this fundamental thesis provisionally, and then it must be clear to you that there must exist some parallelism between the events seen through our senses and the events in the mental reality.

In the previous chapters we observed the cosmos and the CA through our senses only. We call this point of view "the physical aspect" of the universe. It is the world of space and matter. But our worldview would be unbalanced if we were to ignore the mental aspect. In other words, we are obliged to integrate the mind-body problem into our picture.

The first step is an inquiry into the source of our physical world. We obtain our information concerning the external world through the sense organs of sight, hearing, touch, taste, and smell. And, lest we forget, we also use the often ignored sixth sense—the organ that gives us information concerning three-dimensional space. This sense is an organ physiologists have located near the inner ear—the vestibulum and the three semicircular canals.

Sense organs are "feelers." Every separate feeler is sensitive to only one special kind of information. We interpret this information as light, sound, spatial shape, and so forth. In the first year of life we learn how to combine and integrate all these interpretations into a consistent whole.[1] This is what psychologists call *gestalt*—the totality of sense perceptions from which we infer the identity of an object.

Let us first have a closer look at our system of senses: The space-sense is distinguished from the other five senses by its senso-motoric functioning. All the other senses receive the impressions as passive feelers, but the space-sense is put into operation by volition. It is being moved by the bearer in order to excite his impressions of three-dimensional space.

But the space-sense has a crucial feature in common with the other senses. It appears to us as a localizable organ. It is a receptive part, situated in the vestibula-zone, the central cavity near the labyrinth of the inner ear. This cavity contains the receptive organs known as maculae. Maculae are lined along the inner side with hair cells and cilia, hairs steeped in a gelatinous liquid. This liquid contains hard inorganic crystals, called otoliths. These otoliths sink downward, and their pressure bends the cilia. The effect of this pressure is transmitted through the hair cells via synapses, and from the synapses through the special nerves towards the sensoric part of the hearing-sense.

The pressure exerted by the otoliths provides us with information on the position (orientation) of the vertical or Z axis, which is the orientation of the vertical axis of the three-dimensional Euclidean dimensional axes. When we move our heads, the otoliths roll over accordingly and excite other cilia, and this informs us about changes in our position.

Information about the orientation of the other two Euclidean axes, the X axis (to the left and to the right) and the Y axis (forward and backward), is provided through a contiguous, receptive section of the space-sense—through the three semicircular canals (*canales semicircularis*) in the upper part of the labyrinth. The three semicircular tubes are filled with a liquid, the endolymph. This liquid moves slowly with the slightest movement of the head. The three canals are sealed off by a membrane, the cupula, which transmits the pressure-changes of the endolymph to an ampulla outside the canals. This ampulla contains a receptive organ, the cresta, which is built like the maculae but it does not contain otoliths. The hair-cells, and the cilia lining the ampulla, receive the variations of the pressure. This information is transmitted by the cupula through the cilia, the synapses, and then along the nerves towards the sensoric part in the lateral cortex of the brain.

We may learn from this description how we obtain a notion about three-dimensional—and only three dimensional—space—that is, Euclidean space. These three directions are given to us by the sluggish movements of the endolymph in the three perpendicular tubes or canals.

This is no more than an incomplete description, not an explanation. The principal imperfection is that I am caught in a circular argument. I try to prove that we can only have a notion of three-dimensional

Euclidean space by pointing to the three perpendicular canals. But meanwhile I ignore that the description of a sense organ is by itself an interpretation from my sense perceptions of which my notion about a three-dimensional space is a contributory part. Is it not nonsensical to expect a fourth tube giving information about a fourth dimension? The quandary is part and parcel of the mind-body problem we have to solve.

My description of the space-sense suffers from another shortcoming. When I close my eyes and block off all the effects of my visual-sense, I not only receive a notion of the three-dimensional space when I move my head, but I get the same sensation when I do not move my head but move my limbs instead. We know of arthropods (animals with segmented bodies and jointed appendages) that have auxiliary organs of space-sense, with otoliths, in the joints of their limbs.

Not being a biologist I am not familiar with similar organs among the higher animals. But we know that a person who has been blind from birth is able to mold a perfect cube from a lump of plasticine without moving his head. Not only is he not assisted by his sense of touch, but he is even able to "draw" the cube in the air with his fingers.[2]

Our space-sense is an independent faculty. It does not have to rely on information from any of the other five senses. The interaction and the integration of all our sense perceptions into a gestalt is neither taught by parents nor by teachers; it develops in the first year of life. The merging of the space perception with those from our other senses, and especially with our visual-sense, is very intimate.

The amalgamation of our sense of space with our sight has some very interesting aspects. We do not see the external world as a flat picture on our retina, but we feel through our eyes outward into space. It is a subconscious process of coupling our space perception with our perception of vision. This would suggest that the image on the retina would have some importance for the physiologist only, and that we would be permitted to ignore the physical image on the retina and concern ourselves only with the mental process of seeing. But this approach is not justified.

Though we, seeing people, are not aware of the image on the retina, we are nevertheless endowed with the faculty of suggesting a three-dimensional picture by a perspective image on a flat surface—by a "picture." This faculty is also a mystery for me. I may illustrate it with an analogy between the eye-lens on the retina and the lens on a camera, which provides a conical projection of the object onto a flat, light-sensitive plate. I believe that these two processes are comparable—they both belong to the world of our senses. The flat print, the photograph, speaks to us in a spatial language, though the suggestion of three dimensions does not imply that we really perceive it in three-dimensional space. In short, a picture on a two-dimensional surface may suggest a three-dimensional object if it is drawn according to the rules of perspective. It is however impossible to suggest a four-dimensional object with the aid of a three-dimensional sculpture.

A four-dimensional space does not exist in the world of our senses. But we are endowed with yet another faculty with which we are not only able to suggest three-dimensionality with just one two-dimensional picture drawn according to the rules, we are also able to evoke real space perception with a stereoscopic set of two pictures. This sensation is absolutely mental, and nobody has yet invented a physical apparatus to "explain" a mental process. There have been many scientists with materialistic outlooks who have tried to explain sensations, however they have never succeeded.

What really happens when a sensation of space is evoked by stereoscopic means (our acquired faculty to activate our senso-motoric space-sense) is that we fuse in our mind the two pictures into one. This is a much more refined use of the same faculty we acquired in the first year of our lives to fuse the impressions through our left eye with the impressions though our right eye (which are received at a slightly different angle). This is most probably achieved through a simultaneous activation of the two space-sense organs—the one on the right side of the head and the one on the left side.

All the efforts to explain this mental effect by referring it to the physical world of space and matter were doomed to failure. A Dutch physiologist Henri C. Raasveldt wrote a paper on the subject in the *Photogrammetric Engineer*. It was titled "The Stereomodel: How it is Formed and Deformed." It did not bring us a flea-hop closer to a solution. In his first figure he proposes an imaginary third eye—somewhere above the bridge of our nose—that fuses the two pictures. This Cyclopean eye is nothing but an invented physical aid, an imaginary organ meant to evoke the mental hallucination of a space-sensation. It is, by the way, not a hallucination at all, not a pathological symptom, but rather a universal mental phenomenon common to all animals. We even make use of this chimerical feeling to draw very accurate typographic contours—the conventional way of suggesting the three-dimensional shapes of mountains and hills—with the aid of an optical contraption and a set of stereoscopic air-pictures.

During the years I taught this subject, several of my students suffered from slight nausea when they tried to fuse two air-pictures into one image. It is the same unpleasant feeling we get when we overstrain our space-sense. This illustrates the complexity of stereoscopic viewing: Our sight and our space-sense are two independent faculties that we manage in general to merge without difficulty. Under certain circumstances, however, they come into conflict.[3]

Child psychologist Jean Piaget generalized that a newborn does not know a space embodied with objects. All the senses announce themselves separately. He wrote that "to integrate the impressions requires an eighteen-month period of learning."[4]

Our sensation of time is another story. We cannot point to any receptive sense organ, to any feeler, which may be regarded as the provider of our sensation of time. (Let me put here instinct in italics when we refer to time perception, in order to make a distinction between sense perceptions, such as the perception of space, and the feeling of time.) We do not obtain the sensation of time through the interpretation of some sense perception provided by some time-beating source in the external world. This may seem to be a paradox when we consider how often during a day we look at our watches.

We do not always realize that the real source of the perception of time is within us. I tend to refer it to instincts; but the term "instinct" has become a dirty word in psychology. Many mental processes we call instincts function through the reception of special sense perceptions. Many others are not dependent on our senses at all. It is even hard to define what an instinct really is.

The physiologist, who only deals with the effects in the physical world, speaks of the "biological clock." This object, which he may still regard as the bearer of the instinct of time and rhythm, is not larger than a complicated molecule.

Physiologist William Peter Colquhoun[5] described how the long molecular strings of a cell nucleus, the deoxyribonucleic acids (DNA), which bear the heredity of the chromosomes, must duplicate themselves in a strict sequential order, so that atom after atom is being reproduced before the lengthwise division has been completed.

This way of propagation is being performed as if a very accurate clockwork regulates it. Even the tiniest irregularity in this invisible chronometer would lead to a miscarriage.

That this regulating mechanism would be a sense organ is out of the question. Sense organs are composed of numerous cells, and each cell is composed of a very large number of organic molecules. But a unit representing just one single molecule, such as a DNA ribbon, cannot be the bearer of what we conventionally call a sense organ. The in-born chronometer that regulates the duplication of the DNA ribbon must therefore be something else. I do not know what it should be called, but instinct is defective because instincts cannot be defined adequately.

What we may observe within ourselves as a "biological clock" is of a very complex nature—a clock with many hands and different rhythms. The beating of our heart is like a hand of a clock moving past the seconds. Our breath—inhalation/exhalation—beats even slower. Still slower is the daily rhythm of the rise and fall of our body temperature and alternating sleep and awake-states. But none of these rhythms is as accurate as the propagation of the DNA.

All these rhythms would oddly enough not be disturbed in the least if we were to become troglodytes, cutting ourselves off from the diurnal rhythms. This has been shown by many experiments. The same has been found with the biorhythm—the monthly and the annual rhythms.

The idea that our sensation of time should be induced by influences coming from the external world has been ruled out. The sensation of time is a self-regulating internal sensation. The latest discussion between the two camps (the older one represented by Frank A. Brown, and the more recent one by John Palmer and J. Woodland) has been compiled in 1970 under the title *The Biological Clock*. This does not mean that external influences do not exist. They act the modest part of performing small necessary adjustments; for example, when we suffer from jet lag after a flight from London to New York.

But what has been the point of these long circuitous explanations about senses. It is meant to show that our sensation of three-dimensional space and our sensation of time come from two totally different sources. I may even say different in category. The integrating of two kinds of sense perceptions, sight and space, has been found to perform smoothly in our cradles, yet in contrast it turned out to be impossible to integrate our space perception with our instinctive time perception in such a way that the resulting gestalt is in harmony with all possible conditions.

We cannot go beyond the interpretation that objects move or change with a certain velocity in a three-dimensional Euclidean space. It is a system that had been worked out by English physicist Isaac Newton (1642–1727), but it breaks down at very high velocities, and near to strong gravity-fields. We are born and built in such a way that our senses, our instincts, and our intellect are adapted to the forming of a gestalt, enabling us to survive and to propagate.

But how does this gestalt refer to objective reality? We do not doubt that all the sense data, about which we are able to converse because we all receive them, point to elements in a real external world. But who would dare to believe that there is a real identity between our subjectively interpreted essences and the objective existences in this external reality? I repeat: are our senses really meant to reveal the truth of this reality; or are they nothing more but the essential information-gathering tools we use in the struggle for survival. If the latter argument is the case then they provide us with probably much less than the truth, or even with a distorted truth.

Let me illustrate this awkward problem with an example from my own profession. If geologists need information from the hidden depths below their feet they call for the help of geophysicists. Each geophysicist carries an instrument, a "feeler," which is like a sense organ. They also use other instruments, such as a gravimeter to measure the deviations of the earth's gravity fields, a seismometer to discover how sound waves are reflected by an unknown body deep below, or a magnetometer to look for deviations from the general earth magnetic field.

Once they have received all this information from the geophysicists, the geologists trouble their heads what to make of it. What is the "real" thing lying in the depths of the earth? Using geolog-ese they try to integrate all the data into a meaningful gestalt.

For the purpose of this illustration let us imagine the gravimeter told our geologist that there must be something unusually heavy below his feet. The seismometer told from the reflections evoked by artificial sound waves that this anomaly is bell-shaped and that the layering of the adjacent strata is somewhat fuzzy. The magnetometer showed a pronounced magnetic deviation.

But the geologist, at a loss to interpret all the data, decides to drill a test hole in order to take rock-samples— the only information missing. And, behold, it turns out that the "something" responsible for all the strange geophysical data is a very trivial volcanic intrusion that did not make it to the surface. It melted the surrounding layered rocks somewhat and effaced the layered texture. The intruding body is bell-shaped.

How does this example illustrate the activities of our senses? The geologist can put all the geophysical data to the test by taking actual samples form a test-hole. But how are we able to reveal the "reality" behind the data supplied by our senses? Our sense organs present us with riddles similar to the data received by all the geophysics instruments. Just like these instruments they do not give a straight answer concerning the "reality." I told you that from our senses we form a belief that the external world is a world of space and matter; that this impression must be a distortion of the truth—but that it is not a mere hallucination either.

So what is it? The geologists had recourse to drilling-tools. We have to solve the mind-body problem by other means. Mankind has been looked upon as a dualistic creature since biblical times. Consider Genesis 2:17: "And the Lord God formed man of the dust of the ground and breathed into his nostrils the breath of life." But is man really a "ghost in a machine" as English philosopher Gilbert Ryle (1900–1976) put it mockingly? René Descartes reasoned by induction that not only mankind, but the entire cosmos consists of these two separate worlds—a mental one and a physical one. He regarded these two worlds as "two closed systems," each one following its own nexus of causes and effects; but neither interfering in the matters of the other. Descartes was aware that this is a very unrealistic view. How is my mental will able to move my physical hand? How is it possible that my physical brain secretes my mental thoughts? Karl Vogt commented in 1847 on this problem, adding the words: "Just as the liver secretes the bile."

Are these examples not ample evidence that the mental state seems to interfere continuously with the physical, and vice versa? Descartes, and German philosopher and mathematician Baron Gottfried Wilhelm Leibniz (1646–1716) invented weird mechanisms with God as the great magician, to account for these miracles.

Notes

1. This aspect of the development of the mind was studied by Professor Jean Piaget.
2. The space-sense was discovered by the German physiologist A. Riehl in the nineteenth century. Riehl recognized it as an organ that provides information on perceptions of movement ("Bewegungsempfindungen.") The Dutchman Gerard Heymans gives a summary of this discovery in his work *Die Gesetze und Elemente des Wissenschaflichen Denkens*. For further information see H.B. Barlow and J.D. Mollon's, *The Senses*, and A.R. Luria's *The Working Brain*.
3. An explanation of the causes of seasickness and motion sickness has been hypothesized by the so-called "sensory-conflict theory." It is also the accepted explanation for the causes of motion sickness suffered by astronauts. NASA has termed this problem Space Adaptation Syndrome. See Roger B. Swain, "Message from a Heaving Deck," *Discover*, p. 60, vol. 5, no. 8, August 1984.
4. Jean Piaget, *Problems of Genetic Psychology: The Child and Reality.*
5. William Peter Colquhoun, "Explanation and Investigation of Biological Rhythms," *Biological Rhythms and Human Performance.*

5

God, Mind, and Body: Part 2

Let us now look at the mind-body problem from a more reasonable point of view. Imagine a person being observed by a brain specialist. It strikes the specialist immediately that with every turn of thought of the person under observation, a change takes place in the electrochemical brain processes of the thinker. This he observes through his intricate instruments (a positron-emission-tomagraph, for example) and this is conveyed to him though his senses and his interpreting mind—a very complicated chain of processes.

The brain specialist undertakes that the data he or she obtains about the thinking of the patient is of secondary nature. But the patients have no reason to doubt the direct reality of his thinking—thinking is for them a most direct and straightforward reality. Their thoughts are not even conveyed through any of their six senses. The brain specialists, on the other hand, get their information through their instruments and their senses, which are not less selective than the geophysical "feelers" in the geology illustration. They pick up only the kind of information for which they were built.

There must be a real existence behind the object the brain specialists call "brains." But I repeat that the information they receive about these so-called brains and brain processes is so secondary, so indirect, that they cannot make out what the reality of this existence may be.

And now we perform a crucial experiment. We confront the feeling of the thinker with the conclusions of the brain specialist from the secondary information, and then we merge the two aspects into one single event. The following surmise emerges: thoughts look like brain processes whenever they are observed though the instruments and the senses of a brain specialist. Considered from this point of view, brain processes are a kind of shadow from a reality called "thinking." Think carefully before you accept this statement.

If brain processes are no more than so-called sensate interpretations, which are in reality thoughts, we may well descend into the neural systems of animals and into their feelings; for it is not necessary for the patient, whether a person or a dog, to be really conscious of its feelings.

Mental processes, whether conscious or unconscious, may be observed as brain processes or neural process. The new, somewhat more generalized, dogma thus becomes: the feelings of animals look as if they were brain processes whenever brain specialists observe these feelings through their instruments and their senses.

We may descend deeper still into the botanical world and regard the processes the scientist sees as "chemico-physical" as occurring in a mental reality.

I am acquainted with some people who do not share the assumption that a gradual transition between the animate and inanimate world exists. I have reason to differ from this skepticism. We witness how the supposed gap between animate and inanimate becomes more and more irrelevant with the progress of science. Therefore I am convinced that the dogma—"mental processes are the reality. The collateral physical processes are not more than the interpretation from sense data obtained from the mental process"—must be generalized as following: "The external world is in reality a mental one, though it may look like a world of space and matter whenever we view it through our senses."

Brain physiologists of a very lucid mind are often only vaguely aware of this truth. C.U.M. Smith, who wrote *The Brain, Towards an Understanding,* entertained these views:

1. The concept of psycho-physical interaction is a profoundly unsatisfactory theory. (Smith means that there is no simple causal connection between his will to lift his arm and the actual realization of his will.)
2. In principle the peculiar properties of living systems are totally explicable in physico-chemical terms.
3. When Jessica admires the night sky, she is not aware of certain physico-chemical happenings in the occipital cortex. She sees the "floor of heaven." Yet the neurophysiologist would only be able to demonstrate certain action potentials, certain chemical events at synapses.

We may wonder how Smith would integrate these three statements, which tally with my own views, into a consistent philosophical system. Indeed he tries his hand at a convincing generalization, a general inference; that "metaphorically the mental and the physical are the same thing looked at from different positions."

His statement comes very close to the result of my deductions, but the problem in Smith's sentence is the word "metaphorically." Do we avail ourselves of figurative language, or are we trying to come to grips directly with the mind-body problem? Who believes that the fallacy of René Descartes' dualism is a question of "using the wrong metaphors"? On the contrary, Descartes' mistake was caused by erroneous concepts. The first scientist who saw the light was the remarkable Gerard Heymans who wrote *Into the Metaphysical on the Base of Experience*.[1] And another scientist who came close to a similar conclusion was Arthur Stanley Eddington.[2]

Eddington agreed that we are aware of nothing but our own consciousness. He managed, however, to escape from a solipsistic view by pointing at the common character of our sense perceptions, referring to the reality of an external world. As to the essence of this external world he explained, using a neo-Kantian approach, that "it must be a world of consciousness of life—all knowledge of it is knowledge of spirit. The purely objective world is the spiritual world in which physics can do nothing more than shadow or symbolize."

I agree in the main with Eddington's view, but I would prefer to use the word "mind" not "consciousness." Eddington, however, was a notoriously muddled writer and his lucid thoughts were worded in garbled sentences. Susan Stebbing did nothing but yap like a malicious Pomeranian at Eddington's misty utterances. She was mistaken in her pursuit and should at least have made an effort to reduce chaos to order. Eddington's profound truth may then have emerged.

Eddington was a decent Quaker. He regarded as "very rude," Lemaître's term and concept, the big bang, as anticipated from Einstein's theory. It made too many improper and indecent noises, and he said, "since I cannot avoid introducing the question of a beginning, it has seemed to me that the most satisfactory theory would be one which made the beginning not too un-aesthetically abrupt." But this latest digression has no immediate bearing on the subject we are dealing with. So let us return to the mind-body problem.

Several points need further clarification. Firstly it is obvious that the world of mental reality is as a rule a totally unconscious world. Only that very small part of the mental reality, which we observe as brain processes, reflects a conscious reality. This paucity of conscious mental processes is a fact of everyday life. We encountered an example in a long chain of causes and effects in our activated space-sense. We are only aware of space at the moment a brain specialist observes the arrival of the physico-chemical chain in the sensoric part of the brain.

Secondly, there is the point concerning the sensation of vision. Again we become aware of our ability to see only at the last link in the long chain of processes. The brain specialist sees nothing but a long conveyer belt of physico-chemical processes, like light falling on the eyelens, the refraction of the light in the lens, the picture formed on the retina, the process of transmission through the eye nerves, and finally the arrival of these signals at the back of the brain. The seeing person is unconscious of the chain of events; but suddenly something happens of which the person is clearly aware and the brain specialist is not—he sees and he knows that he sees.

When we think about this marvel, we might forget that the process of seeing is an unconscious mental chain, and that we may thank heaven that we are unaware of the intricacies of the process. Life would be unbearably complicated if we would be conscious of all the links in the chain the physiologist perceives as physico-chemical. Nature is mental—but wisely parsimonious when it comes to consciousness.

The third point is that we must, in all the explanations, make a very sharp distinction between thinking as a conscious mental process and the contents of the thought. We are rightly convinced that we are engaged in the act of thinking though we may be mistaken about the correctness of our thought.

The final point refers to pain. Pain is a feeling. Feelings belong to the mental world and the mental world is a reality. A doctor stands at your side and asks you a simple question: "Where do you feel pain?" Both you and the doctor ignore the heterogeneity of the components in this question. Both the location of the pain and your suffering are notions borrowed from the physical world of space and matter. But your pain is a sensation, hence mental reality.

The perceptions of a baby are slightly different. It may feel the same pain but it has no notion about the location, nor in fact of its body. It is still one with the universe. The raw information from its senses is not yet digested by its mind—there is no construction of a gestalt.

At the end of its first year, the feeling of individuality is as yet undeveloped. The toddler takes a stick in her hand, but she does not feel it as an inanimate instrument. It is simply the extension of her arm. The point where her body passes into the external world is still a bit vague. Finally she finds it and notices that her finger feels the stick, but that the stick feels nothing. This illustrates how our mental world and the world of our senses become thoroughly intertwined in the course of our psychogenesis and the maturing of our mind.

That the world of space and matter is a mental deduction, the interpretation of our sense perception from a mental reality, is the principle constituting the fundamental base of the Idealist School. Baruch de Spinoza was the school's pioneer and his principles were handed down via Gottfried Leibniz and German idealist George Hegel (1770–1831) to the other German thinkers of the nineteenth century. Finally Gerard Heymans presented them in rational scientific form in his book *Into the Metaphysical on the Base of Experience.*

Absolute idealism, a special brand of the idealistic view, teaches that the world of mental reality should be identified with God—also called the Absolute. This view is expressed in Spinoza's slogan: *Deus sive Natura?* It seems to me that this was also once the view of Albert Einstein.[3]

We are compelled to reject the Absolute Idealistic view. That the world of space and matter has been created indicates that God the Creator cannot be identical with the physical world. But we have just completed our deduction that the physical world, the world of our senses, is more than a shadow of our mental reality. Our feeling of modesty, our notion that our senses are not ideal feelers and are deficient in their perception of everything the external mental reality might contain, may leave open the possibility that there *is* something in the mental reality. There is in the mental reality something our senses cannot perceive, something that has not been created. I am nonetheless convinced that not only our physical perceptions but also the mental reality has been created out of nothing.

I make this deduction because of the omnipresence of time. Time is a vital parameter in both worlds—in the physical shadow as well as in the world of mental reality. Thoughts without time are unthinkable because thinking is a process. And time, which has been created, elapses in both worlds. If it was created in the physical world it must also have been created in the mental world. An eternal past is as absurd in the mental reality as it is in the physical world.

How did the views on space and time develop at the beginning of the twentieth century (at about the same time Gerard Heymans wrote his work on metaphysics, but before Einstein transformed the world of physics beyond recognition). Two figures stand out in this short period of pre-relativistic philosophy—F.H. Bradley and J.E.M. McTaggart. Bradley, who lived just at the turn of the century, ran into difficulty because of his peculiar view on essences.

"Properties," so he said, "relate to a perceived object, but why do they not relate one to the other?" For example: the properties of sweet, white, hard, and crystalline relate to sugar; but sweetness, whiteness, hardness, and crystalline do not relate one to the other.

It must certainly have dawned on Bradley that the limitation about which he was talking belonged to the brute facts of how the various information received though our senses announce themselves to our mind. In modern terms, a gestalt is related to its components that do not need to be interrelated. We are not able to think and talk in other terms.

But Bradley repeated continuously that a judgment must have diversity in order not to become a tautology—yet the judgment must unite the diversity. Bradley deemed this to be impossible, and he finally concluded that all our notions of space, time, and causality are defective. This made him doubt the reality of all these parameters, including time. This view is called a personal Idealistic Attitude. I contend that behind the notion of space, perceived through our space-sense, there lurks a mental reality. The true character, however, cannot be known because our space-sense is the only instrument through which we perceive just that which we call space. The sensation of time is however a direct one because we perceive them directly through our mental faculties. This is why I am bound to reject Bradley's pessimistic view that our perceptions are hallucinations.

On God he commented thus: "If you identify the Absolute with God, that is not the God of religion." And he reasons that the God of religion is a personality. Hence God can only be an appearance in Persona of the Absolute. Hegel defined the Absolute as the Absolute Spirit, and he added that there cannot be anything outside spirit. Bradley agreed, but we have to reject Hegel. The spiritual, i.e., mental, world has been created out of Nothing, by the Nothing, and God stands aloof from His creation.

Let us now turn to McTaggart, whose ideas on time were no less weird than Bradley's theory that both space and time are hallucinations. Here are the main points of his reasoning.[4]

He discerned an A series, marking past, present, and future, from a B series referring only to before and after. He maintained that the A series implies changes but that the B series refers to stationary items on the calendar. "When we say," so he explained, "about the events in the A series, that they have been past, are present, and will be future, it must mean that an event is present at a dimensionless moment; at a moment thereafter it becomes past; and we may choose a moment earlier when the event was still in the future." "Is" here means a static and unchanging condition. And then followed his sophism that time is not real; that time is a kind of being and not a becoming; and that time only seems to us to be real through a third, C, series—the timeless growth of perceived reality, including the present and the past.

The whole system reminds us of Aristotle, and the sophism is the spurious reasoning of the Greek philosopher Zeno of Elea (c. 450 B.C), who tried to bamboozle his audience with the example of Achilles who could never catch up with the tortoise because this fast runner could only place his foot on the spot where the tortoise had been before. This mistake is characteristic of people who have no idea of measuring. We measure length in finite distances not in infinitesimal points. When we measure the speed of Achilles by the number of his steps per minute— the length of his step being known (whereas the speed of the tortoise is known as well) and after we gave the tortoise a decent start, a certain moment must arrive when the distance between one of Achilles' feet and the tortoise has become shorter than the length of one of his steps. The following step or the one after must become an overstep beyond the tortoise and Achilles has won the race.

Shortly after, Einstein appeared and turned space and time into space-time. That four-dimensional space-time is neither imaginable nor depictive was not the fault of his theory, but a defect of our fallible senses. No sane person would try to improve the faculties of our limited imagination.

That space-time in Einstein's equation is a continuum is much worse. A continuum is a medium that may endlessly be divided—any finite volume consisting of an infinite quantity of infinitesimal dimensionless space points. A continuum is an example of the actual infinite, an absurdity. Einstein knew this, but I doubt he ever seriously tried to solve the problem. He entered the twentieth century with the same trepidation as most of his colleagues, and exactly on this point he did not see eye to eye with several among them— especially not with the Danish physicist Niels Bohr (1885–1962). He was ready to give up a strict causality. "God does not play dice," was Einstein's reaction to the new quanta-mechanics and the principle that the sub-atomic world may follow certain rules but is in principle unpredictable. A theory that fails to be causalistic must be an incomplete theory, so he said.

Maybe Einstein had a point here, but he meant an "incompleteness," which is not the same as the "incompleteness" I have in mind. Einstein wrote on 24 May 1940 in *Science*: "Science is the attempt to make the chaotic diversity of our sense-experience correspond to a logically uniform system of thought." Science and physics are "sense-minded"; and sense perceptions have the last word.

We are not, however, dealing with science but with philosophy and metaphysics. I explained that science, the world of the physicist, can never find the objective truth of the external world because it takes the world of our senses for granted. We have seen that the world of our senses is a shadow-world. The movements of shadows follow certain patterns but are less predictable than the movements of the real objects which cast the shadows. We have deduced that the reality of these objects is a mental one. We may guess that the behavior of the mental objects is strictly causalistic, but that our inadequate minds cannot help looking at their shadows with the arbitrary movements. Our senses and our minds may never crack the secret.

Should we ever be allowed to have an unmediated view on the strictly causalistic mental reality behind the freakish behavior of subatomic particles of the world of the physicists, we may also expect that we may be allowed to catch a glimpse of the reality behind the physical aspect of the *creatio ex nihilo*—a glance at God's own kitchen. Even the very idea seems to be a blasphemy, but I am aware that I am only expressing personal unreasoned feelings. We cannot predict the progress of philosophy.

Meanwhile physicists are hard at work looking for a solution to the question of how Einstein's space-time may obey the law that the actual infinite does not exist. They try to make space-time as atomistic as matter. Let us take the atom, the smallest particle of a chemical element, as a model. The essence of its subatomic components must differ from the essence of a chemical element, otherwise the atom would not be its smallest particle. It has been established that even the properties of matter in general—that is, its having mass and its occupying space, lose the absolute meaning in the subatomic world. We are, nonetheless, able to talk some sense about this micro-world, a sense derived from indirect experiences, direct perception of subatomic particles, and even atoms as being out of the question.

This is as far as science dealing with the smallest parts of matter can go. Talking about atoms of space-time is much more awkward. In analogy with atoms of matter, the components of space-time atoms must have a non-space-time character. The great problem is that even the space-time continuum can neither be imagined nor depicted. It falls, due to our defective experience, apart into space and time. We may figure a non-spatial reality, the mental reality; but a reality without time is nonsense. It disappears into nothingness. Time is a mental reality. It is not less hard to imagine atoms of time—their components bearing the character of non-time.

Roger Penrose of Oxford proposed that space-time becomes foamy and unpredictable on a subatomic scale, and that the breakdown of conventional geometry begins as early as the diameter of the atomic nucleus. He embarked upon his "twister" project about subatomic space-particles, following the laws of a "pre-geometry" of John Wheeler. All of this paperwork carried out in the study hides tricky pitfalls for those not yet versed in philosophy. This may follow when we take stock of the discoveries that the big bang is the beginning of the past and has become an unassailable truth. If the influence of gravity is strong enough to overcome the explosive force of the big bang after it has been spent or not, it would decide whether the universe in the far future would contract again into nothingness. It is this prospect—dubbed the big crunch—which is being contested by the majority of physicists. The density of matter in the universe seems to be too low and it is very doubtful if the mutual attraction will lead to a general contraction.

Probably the universe is "open," which means that it is "semi-infinite" in time. It has a finite past and a potential infinite future. This has something in common with religious articles of faith that the human soul is born but has an eternal life.

But let us assume an occasionally proposed possibility of an "oscillating universe."[5] It is the hypotheses that the universe alternately crunches itself down and explodes again rhythmically in an endless oscillation. Scientists have, of course, no real evidence. The idea has, moreover, very serious flaws from a philosophical point of view. We are not allowed to introduce actual infinity through a back door. To believe in an oscillating universe means that we have to align it with the fact that time has a beginning. But when should time begin? Should time begin at some arbitrary moment when the size of the universe was at a minimum? It is not much more reasonable to accept that the "first moment" of time coincided with the one and only big bang. It seems to me that an oscillating universe is a semi-intellectual entertainment. One of the philosophical problems would be, for example, how to view "timeless intervals between a big crunch and a new big bang." An interval must have a duration.

A really awkward proposal is to associate the arrow of time with the alternating contraction and expansion of the universe. This has been the hobby of Thomas Gold.[6] Gold suggested that the universe oscillates and time reverses not in the fireball but in the point of maximum expansion. It would imply that time would come to a standstill and would run backward as soon as contraction sets in. Roger Penrose of Oxford deemed it impossible. He argued that the arrow of time, or the direction of its flow, is fixed by gravity and that the clustering tendency of gravity persists—that it does not change whether the universe is expanding or contracting. In other words, time is irreversible in principle.

I have a few additional questions concerning Thomas Gold's proposal. I fail to understand how the universe should persist in its existence when time stops. What would be the fate of all the other physical parameters? Would time coming to a halt not imply that we would be saddled with a timeless universe? We have already answered this question sufficiently. A timeless universe cannot exist neither in the gestalt of a physical world, nor as a timeless, mental reality.

And thus we summarize that a mental reality has been created only once—out of nothing. This nothing, the first cause of creation, transcends the mental world it created. The Creator is neither an empty void nor mind or spirit, and the more essence we deny the Creator the more evident (existent) the Creator becomes. The Creator *is* existence—completely—and has no essence.

Bertrand Russell persisted with his denial of God's existence until his death in 1970. I believe his skepticism was against his own better judgment. He stuck to his guns until the barrels became an incongruous part of his very personality. It did not help to rub his nose into observed facts or logical deductions. In his letter, "Why I am Not a Christian," he maintained: "If everything must have a cause then God must have a cause. If there can be anything without a cause, it may just as well be the world as God, so that there cannot be any validity in the argument."

Is this a valid argument against God's existence? Of course not; and I counter Russell. Leucippus the Greek allegedly said that "nothing happens without a cause and every event happens of necessity." This principle of universal causation is so self-evident that it has not even been dented by the modern physics of the microcosmos, where more than one alternative effect may be engendered by one and the same cause. Spontaneity does not exist. It is misused as a subterfuge by dullards who fail to find the causes, or by children who lie to their parents with explanations like, "the window pane just broke."

Every change must have its cause. The premise misused by Russell that everything must have a cause is erroneous; for that would imply the absurdity that the First Cause, which is a thing too, must have a cause. And thus we have to maintain that the First Cause, known as God, engendered the first change—the Creation of the world out of Nothing. But the real trouble with Russell is, of course, not that he would fail to understand that the question whether a First Cause has a cause is absurd, rather it is that he did not believe in the existence of a First Cause.

But even believers in the existence of God may fall into the trap of confusing sophisms. They are of a different kind. We established, through the vindication of the CA, that the OA and the AFD, which rely on the nonexistent essences of God, are invalid. The rejection of positive attributes settles once and forever the paradox of God's alleged omnipotence.

We find the following questions on the Creator listed in Peter A. Angeles's *Dictionary of Philosophy*:

1. Can He create a square circle?
2. Can He undo the past?
3. Can He create a rock big enough so that He cannot move it?
4. Can He invent problems that He cannot solve.
5. Can He annihilate Himself and never come back to life?
6. Can He deny His essence.

In whatever direction we twist and turn these questions and try to solve them there are always two nonexistent attributes that try to creep in—God's omnipotence and God's omniscience. Omnipotence is actual infinite power and omniscience is actual infinite knowledge. Can an actual infinite not existing in Creation be attributed to its Creator? I believe that this is a senseless question because the Creator must always remain the essence-less Nothing in the eyes of His creatures.

When we maintain that He created everything we mean all the things we know about and all we do not know about. The six questions I quoted above—and others like them—are nothing but products of sick imagination. I agree that hallucinations are real and existing thoughts, but this does not mean that the contents of all real thoughts are reality. And this implies that we are allowed to relegate the six questions to the nonexistents.

Another common sophism is the attitude of the Materialists. They dare to claim that the world of space and matter is exactly mirrored by our senses. They therefore call our sense perceptions reality and regard mind as a hollow fiction. Did it ever occur to them that without the existence of our mind—which they call fictitious—the whole world of space and matter would be unknown; and that it—from their point of view—would not even exist? For no perception of the so-called world of space and matter is possible without our senses and the mental faculty to interpret what we perceive.

Modern philosophy engendered several variations of this fallacious materialism. One of them is the "Identity Theory." These views are bundled by V.C. Chapell and I picked out the crucial paper by U. T. Place, "Is Consciousness a Brain Process." He looked at the mind-body problem through the wrong end of the telescope. He believed that consciousness is empirically identical with a brain process in the same way that a cloud turns out to be identical with a mass of tiny droplets suspended in the air, or that what we experience as lightning is "nothing but" a motion of electrical charges. This view is a half-truth, and therefore an untruth.

Consciousness cannot be identical with a brain-process because it is a direct experience. On the contrary, a brain-process is only in a certain way identical with consciousness. We stated that consciousness observed though instruments and senses look like brain-processes. Consciousness is reality but brain-processes are its "shadow."

The identity theory is an odd, old materialistic view in a modern disguise. It denies the reality of consciousness, which is only a brain process. I regard behaviorism as the backward and benighted child of the identity theory. John B. Watson and B.F. Skinner tried to buttress this flimsy construction as follows: "What we called thinking is nothing but involuntary movement of our vocal cords, a soundless talking to ourselves, contractions of muscles, and reactions in the skin. Emotions do not exist. They are just processes in the bowels."

I am acquainted with a professional viola player who could not control the wiggling of the great toe of her right foot, an arbitrary twisting and turning during her performance. Would Watson conclude that her musical emotions are only imaginary and nothing but movements of her toe? Or would it not be more to the point to conclude that these movements, the more outspoken the deeper she is involved in her music, are the psychosomatic expressions of her mental state?

The deep error of the behaviorists may be the consequence of a certain vagueness where we should disregard essences when an existence must be defined. Do accidental attributes still belong to an existence? The predicament becomes obvious when side effects are overrated; and

Watson and Skinner are talking about nothing else but side effects, such as the uncontrollable movements of that toe. The clucking noise of the larynx is a side effect of drinking; but drinking is in reality a physical process, the quenching of thirst, the relief of an unpleasant feeling.

Behaviorists mistake side effects in the physical shadow-world for essences of realities, and this serves only to compound the error.

Notes

1. *Einfuhrung in die Metaphysik auf Grundlage der Erfahrung.* My exposition is a paraphrase of the relevant passages in the second part of Heymans's work.
2. See Susan L. Stebbing, *Philosophy and the Physicists.* And for a short compendium see J. Passmore, *A Hundred Years of Philosophy.*
3. T.L.S. Sprigge, *The Vindication of Absolute Idealism.*
4. From his two volumes, *The Nature of Existence*, and his first paper, "Mind," in 1908.
5. See Nigel Calder, *Einstein's Universe*, p.218.
6. See Timothy Ferris, *The Red Limit.* Chapter 10, "The Expanding Universe of the Mind."

6

The Spanish Intermezzo

When we take the forgoing into account, it appears as if the Jewish thinkers of the Middle Ages had a lighter burden than their Christian colleagues, since they did not have to bother about moral articles of faith. But a closer look reveals that the tasks of the Jewish philosophers was not enviable in the least. The Pauline doctrines were forerunners of neo-Platonic Christian mysticism, and it was exactly neo-Platonism that Judaism had to reject. It was forced to embrace the alternative, the philosophy of Aristotle. But this brought about the same difficulty with which Christianity could not cope. A head-on collision with Genesis and Creation, which cannot be reconciled with the eternal world of Greek philosophy. I cannot read from their writing that their long and elaborate reasoning brought them one step nearer to the solving of this contradiction. (The most famous Jewish philosophers of this period happened to live in Spain.)

Maimonides opened his doctrine of God's attributes in his standard work *More' Nevukhim* in the usual way of his times. He took God's existence for granted and he used Him as his point of departure. He simply ignored the soundness of the CA of Saadyah Gaon, from which one is compelled to deduce that the *Creatio ex Nihilo* cannot be the bearer of attributes which He created. I believe that Maimonides committed this mistake as a consequence of his unreasonable reverence for Aristotle who tried to convince us that the impossibility of the actual infinite could not be applied on time.

But Maimonides quoted two other philosophers, Ibn Sina (980–1037) and Abraham Ibn Daud of Teledo (died 1180) who adduced their special argument of God's existence. Their reasoning has been translated by F. Bamberger as follows: "If there is a being which is subject to becoming there must be with the same necessity a being which is not subject to becoming. This being, which cannot cease to exist or to fade away, is for that reason a being of necessary existence.

One of the first objections that may cross our mind is that in this world all the effects are the necessary consequences of causes. But Bamberger had an answer: "The transient becomes a necessity through its cause." This means that its necessity is a limited one, a relative one. Its existence is only secured through the possibility of its being, and through its non-being by the addition of the cause. The really necessary being, on the other hand, has a necessity according to its nature, which means that it attains its existence through its essence. But he adds this being, which is not subject to becoming, must be an absolutely simple one; this because they believed that a plurality in a being would imply a becoming plural. The elements must have been united by a higher cause, which is the Simple Being.

Julius Guttmann, who wrote *Philosophies of Judaism*, made some comments very much to the point. "We must concede that the demand for a congruence of essence and existence clearly contains the main element of (Anselm's) OA."[1] But there is a difference between the standpoint of Anselm and Ibn Sina. Anselm endeavors to establish God's existence by a (spurious) ontological deductions, while Ibn Sina and Ibn Daud used God's existence as a postulate, an article of faith. It should, however, be clear that neither the postulated God of Ibn Sina and Ibn Daud, nor the ontologically expounded God of Anselm, have any validity as a scientific proof of God's existence. Thus, we have to return to our well-established CA, the First Cause. We found, indeed, that it is a grave mistake to deduce existence from essence.

This laborious elaboration boils down to the following conclusions: Maimonides believed rightly in the simplicity of God and in the absence of positive divine attributes, but his conviction was based on the wrong argument. God must be simple in His way because Zero, the Nothing, cannot be plural. God must be devoid of positive attributes because He created them, and attributes He created are not His attributes—simple and clear. But Maimonides did not care to listen to Saadyah Gaon, and he followed Ibn Sina and Ibn Daud instead.

The other inconsistency is that he believed in God as the First Cause in defiance of Aristotle. Why was he not able to content himself with Aristotle's "unmoved mover?" The reason is that an unmoved mover cannot be God, rather it is, at best, the evoker of movement. He cannot be the creator of the moving object because this is an act we may only expect from God. This explains why Maimonides willy-nilly departed from Aristotle. The old philosopher's religious conviction took preference over Greek philosophy.

Maimonides had some difficulty addressing an attributeless God. He tried to solve this problem by introducing attributes through the back door. Because positive attributes were rejected he had to invent other kinds that had to be reconciled with the First Cause and with the host of the Lord's attributes extolled in the Bible. He tried out three different attributes: (1) attributes of negation; (2) attributes of identity; and (3) attributes of action. Let us see what they meant and if they are acceptable.

Maimonides had saddled himself with an additional difficulty. His aim was not just to describe God with his new attributes. He also wanted to learn something more about God with the aid of these attributes. Small wonder that his endeavors did not fare well. But the aim of a philosopher is not the aim of a pious believer. His aim is first to find God and, second, to be found by God.

Let us first try to find out if we may discover something new about God with the aid of the attributes of negation, which are negations of positive attributes—they state what God is not. Maimonides takes the example of somebody who has no idea what a ship is and he tries to impart the knowledge to him by telling him what a ship is not. So he crosses out one by one all the negative attributes until the questioner at the end may hit on the idea.

Brian Davies in *An Introduction to the Philosophy of Religion* shows that this method of argumentation can never be exhaustive enough. "It is simply unreasonable to say that someone who has all the negations mentioned in it has almost arrived at the correct notion of a ship. He could equally be thinking of a wardrobe."

But I have a still more convincing objection. The only adequate idea about God is that He is nothing knowable in any terms of all the ideas expressed in a human vocabulary, which means that we are unable to define Him in human language. We may cross out all the attributes one after the other and finally get at the very last attribute on the day of judgment and indeed "be nearer my God to thee," but it would not bring us a step nearer to our goal. We would not have gathered any new knowledge about God.

The second type of attribute, the attribute of identity, is meant to imply that God does not have attributes, but that He is identical with His attributes. He is not defined as being Almighty, Omniscient, Bountiful; but He is Omniscience, He is Omnipotence, and He is all Bountifulness. We do not change the kind of meaning of an attribute when we raise it to the superlative. Omniscience is a kind of science; Omnipotence is a kind of power; and all Bountifulness is still a kind of goodness.

If God created all the attributes, He also created their superlative. If God cannot be the bearer of that attribute He created, He cannot be the bearer of the superlatives either, and still less be identical with these superlatives. If God would be identical with the attributes He created Himself, He would have created Himself.

And finally the third type—God's attributes of action. Creating is action par excellence and there is not a bit of doubt that the Creator has the faculty of creating. The following reasoning has two aims. We have to find out if we have the right to call the Creator's faculty of creating an attribute of action and secondly if the so-called attributes of action do teach us something new about God if we study them one by one.

Maimonides could not have been aware of his mistakes, and we cannot blame him that he could not see seven centuries ahead. That he was mistaken follows from a valuable contribution of Immanuel Kant to philosophy, a new insight into attributes.

A sentence in which the predicate is implied in the subject does not tell us anything new. Kant called it an analytical judgment. Only sentences in which the predicate is not implied in the subject tell us something we did not yet know. Kant called them "synthetical judgments." For example: the sentence "my boyfriend is a human being" does not tell anything new because boyfriends are human beings. It is an analytical judgment. But "apples ripen in Autumn" may be regarded as a synthetical judgment, unless we have already implied the time of ripening in the definition of an apple. I may, as another example, tell that "your boyfriend repairs watches" and I have added a new attribute of action; it is a synthetical judgment. That he plays the flute is still another synthetical judgment. And so I may go on ad nauseam enumerating the multifarious activities of your boyfriend in a long row of synthetical judgments.

It may seem at first sight that we may break up God's activity as a Creator into a myriad of creations of separate items, and ascribe to Him a myriad of attributes of action. This is however an obvious sophism. They are all analytical judgments; they do not tell us anything new about God; that is God defined as the First Cause. To make it still worse for Maimonides, it is at least very doubtful if we have the right to call a so-called attribute an attribute if it appears in an analytical judgment. An attribute may be called an attribute only if it tells something new about the subject. Maimonides doctrine of attributes collapses like a stack of tin cans in a supermarket when we view it in the light of modern philosophy.

Let us turn to Maimonides' views on moral sense. He made a very valuable discovery: Good and bad have nothing in common with correct and false. We have only to restrain our mind to follow him because of the weird vocabulary he used. His term "necessity" means logical soundness. Maimonides said: "With regard to what is of necessity, there is no good or evil at all only the false and the true."[2]

It is to Maimonides' credit that he departed on this subject from the old rules laid down by all his predecessors, especially by the Greek moralists, who believed that a good man is guided by his "adequate" thoughts. Maimonides was probably the first who intimated that our common sense and our moral sense are worlds apart. But it is regrettable that he was too steeped in medieval superstitions to stick to the point. He parroted the old doctrine that the human mind has five domains:

1. The nutritive part, or the power of attracting, retaining, digesting, excreting, growing, and procreation of the kind;
2. The sentient part, or the power of the senses;
3. The imaginative part of our memory;
4. The appetitive part, or the domain or our tastes and inclinations; and,
5. The rational part conferring more or less with our ego.

He did not allocate a special domain to our moral sense, though his discovery about moral sense should have compelled him to do so. This omission was the main cause of his confusion. Here follows his definition of his fifth domain: "The rational part is the power found in man by which he perceives intangibles, deliberates, acquires the sciences, and distinguishes between base and noble actions."

This definition is defective, even according to Maimonides' standards. Being able to distinguish between base and noble actions is a moral faculty, but Maimonides only found out that our moral faculty and our faculty of reason must be located in two different domains. In modern language, he bound ego and superego together.

His confusion deepened further on: "Know that the disobedience and obedience of the law are found only in two parts of the soul, the sentient part and the appetitive part." He must have been unaware that he contradicted himself. Should the rational part indeed contain the faculty to distinguish between base and noble actions, we would have expected that it must also contain the faculty to obey the law, which would mean that the sentient part and the appetitive part cannot be the only parts in which the disobedience and obedience of the law are found.

The definition of the sentient part given by Maimonides does not even justify to regard it on a level with the other four parts. It is the total of senses, feelers receiving and acquiring the raw information from the external world. That our sense perceptions may or may not play their part in a moral decision depends on the circumstances calling for a moral decision.

Finally, the appetitive part, defined as the part by which a man desires or is repulsed, is nothing but the total of animal propensities. Moral sense on the other hand is not the total of our drives and wills, but our faculty to choose the one among our drives and wills compatible with the principle of moral behavior. Our choice has to tally with Heymans's principle of objectivity—that one has to regard one's own interests on a level with the interests of our fellow men.

Maimonides revealed his garbled reasoning especially in his arguing with an opponent. In the first chapter of his *More' Nevukhim—The Guide to the Perplexed*, he tries to counter an imaginary objector who says, "It is manifest from the clear sense of the biblical text that the original purpose with regard to man is that he should be as the other animals are, devoid of intellect, of thought, and of the capacity to distinguish between good and evil."

Maimonides corrected his opponent: "Adam and Eve were not devoid of intellect, but of moral sense." This is of course correct, but he concluded his reply in utter confusion. He tried in vain to prove that the conventional interpretation of the episode is correct and that Adam and Eve are nonetheless accountable for their transgression. It never occurred to him that this transgression, by which the couple acquired their moral sense, had turned the two into creatures created in the image of God. From two amoral beings they became moral people.

These confusing ideas are bad enough, but not as bad as Maimonides' most bewildering views of moral virtue. Aristotle said that in the context of his ultimate aim to know God, the highest degree of knowledge could only be achieved by our *nous poietikos*, our active intellect.

Maimonides regarded this aim as the only aspiration that makes life worth living, because of his conviction that the immortality of the soul depends on knowledge of God. Julius Guttman, author of *Philosophies of Judaism* describes this view as follows: "Immortality is a necessary consequence of the level of knowledge reached. It is thus intellectually and not morally determined. Moral perfection serves an ultimately intellectual end in itself."

This is a perniciously demoralizing attitude that pervaded the minds of so many among the medieval philosophers. Among the Christian thinkers this attitude was a matter of course. The highest level is not moral virtue but faith. So why should the intellect not interpose itself between moral virtue and faith? But the rabbis rejected this value scale out of hand. They reproached Maimonides for his subjugation of moral virtue, which became a servant of an intellectual end. Although Maimonides was a pious Jew, there was no place in his philosophy for *Limud Torah le'shema*, which means "Torah study for its own sake." And this was regarded as a *lèse majesté* committed against the sovereignty of moral virtue; the very pinnacle of the human soul. Maimonides' ways were, in this respect, un-Jewish.

It may be said to Maimonides' credit that he was not too obstinate in his refusal to regard moral perfection as an end. He even contradicted his own words more than once and as the author of *The Guide to the Perplexed* was himself sometimes a guide in quandary.[3] Maimonides maintains in part 3, chapter 54 of his guide an opinion on morals that cries to Heaven. He calls a man "of the very highest perfection," his so-called fourth species—not the man endowed with the highest moral virtue, but the man who has acquired the highest rational virtues, in teaching himself correct opinions concerning divine things. He values the morally perfect as standing on the lower level; such a man belongs to the third species. "Moral perfection is," Maimonides said, "only a preparation for something else and not an end in itself."

And then he went about looking for a good Bible quotation to prove that he was right, and he claimed to have found one—Jeremiah 9:23–24: "Let not the wise man glory in his wisdom…but let him that glorieth glory in this, that he understandeth and knoweth me, that I am the Lord which exercises loving kindness, judgement, and righteousness in the earth; for in these things I delight, saith the Lord."

This means nothing else but the obligation to heed the dedications of our moral sense. The three virtues I underlined in the Bible quotation stress that good works are God's delight, not human wisdom—a clear message that is the exact opposite of Maimonides' aim. One may ask why I spend so many words on Maimonides' teachings while the harvest is so lean. I do it because so many sages of our days, including Jews, reveal an unreasonable adoration as if they were endowed with divine qualities of wisdom. Maimonides' merits lay in another domain. We, living in a different world, do not believe anymore that he acquired an immortal soul among the dead on account of his monumental codification of the Talmud, and of his insight in psychosomatic disease.

Levi ben Gerson, alias Gersonides (1288–1344) went even a step farther down in depreciation of moral virtue. However, we will not understand his view on morals without at least a brief description of his general philosophical outlook.

Gersonides was one of the very few medievals who had the courage to get to the bottom of the Kalam CA, and who understood its stringent validity. He rejected Aristotle's view on time—time has no exceptional position among the other parameters, and it obeys the law that the actual infinite does not exist. There is no eternity. But this is about all the honor I am ready to grant him. It sounds absurd that he did not accept the idea that God created the world out of nothing, with empty hands. There must have been, so he mused, a *"prote' ousia,"* a primeval matter, out of which the Supreme Thought must have molded the universe. It is an idea he borrowed directly from Aristotle. The "Prime Mover," which is the "Supreme Thought," thinks in terms of *eidos* and stands above *hyle*, the hylic world out of the *prote' ousia*. Who else could have created it but God? For Aristotle this was not a problem at all, since he believed the world has an eternal past. But it certainly must have constituted a problem for Gersonides who accepted both the narration in Genesis and the validity of the Kalam CA—the creation of the world out of nothing.

Gersonides decried the sophism of Maimonides, that a multitude of attributes would cut the unity of God's essence into bits. The multitude of properties of this one apple does not split this one apple into a multitude of apples. Though this is correct reasoning, Gersonides overlooked that it is not a valid argument to ascribe attributes to God who created all the attributes. Gersonides' medieval idea of God was to a great extent borrowed from Aristotle. "Supreme thought," which can only think with the thought of His "active intellect" along "lines of induction," of generalizations, was the highest level of thought Gersonides could imagine. God thinks in genera; thinking in species is a human imperfection.

But the wise and the sage are capable of imitating God in their way of thinking. Maimonides understood the great difference between God's way of thinking and the human way of thinking, but Gersonides was not so modest and humble. He even maintained that everything that is absurd according to human judgment, must also be absurd in the eyes of God.

This difference in outlook had some weird consequences: Maimonides called himself an enemy of astrology, because the ways of

God are not the ways of the mortals, and that is why divine thoughts have no connection with sub-lunar events. Gersonides, on the other hand, entertained a daring idea, we might even say an Einsteinian one, that the laws of nature are universal. It is of course a great question if this universality of the laws of nature is also valid in God's own realm. But Gersonides believed that this is indeed the case, hence he concluded that there must be a direct interrelation between the divine will and the divine decisions, and the actual happenings in the sub-lunar world. The human mind has, however, the good fortune to be able to elude fate by its own endeavors—though only under the condition that it is aware of what is in store. To acquire this precious piece of information one has to consult an astrologer.

Though Gersonides did not accept Aristotle's view on time, the old Greek always remained an object of his unreasonable admiration. The consequence was, of course, the typical medieval dilemma of how to reconcile the two incompatibles: Aristotle's philosophy with the Bible. The title of Gersonides' magnum opus, *Milchamot Adonai* (The Wars of the Lord), may symbolize his own inner conflict how to maintain his dual loyalty.

But he inflicted the most outrageous insult on the virtuous and the pious with his special brand of immortality. The Aristotelian view that the soul, the psyche, is as mortal as the body, may be an eye-opener for many an ignorant clergyman. Aristotle pointed out that according to everyday experience the faculties of seeing and hearing die away with the dying person. The act of seeing cannot exist without the living eye. Aristotle regarded the faculty of the senses as being part and parcel of the *eidos*. This concept is expressed in the tenet that, "the soul psyche is the form *eidos* of the body *hyle*." Soul and body are interdependent—the death of the one entails the death of the other.

What Aristotle regarded as immortal was not the faculty of the senses, but the knowledge acquired through the intervention of the "active intellect." This was also the conviction of Maimonides. But Gersonides was a hair-splitter. He was not satisfied with the statement that the immortal part of the soul is the part acquiring knowledge; but he regarded this only as the general idea that had to be thrashed out in detail. He agreed with Aristotle that the passive intellect cannot be considered as a serious candidate for immortality.

St. Augustine had written about sense-perceptions a thousand years earlier in his *Confession*. He wrote: "—for this also have the horse, and mule, for they also receive through the body." But Gersonides did not only consult Aristotle for his final judgment; he called in the opinion of another odd couple—Alexander of Aphrodisias (about the year 200) and Themostitos (317–387). These two were peripatetic philosophers, which means that they roused their inspiration by pacing up and down. Alexander believed that the *nous pathetikos* dies with the body and Themostitos believed that the *nous pathetikos* leaves the body.

But we shall not descend into Gersonides's motives why he decided in favor of Alexander. More important is the problem what should be understood by the term "acquired knowledge." Certainly it does not mean only memory. This was also the opinion of Thomas Aquinas: the dead horse of Joan is nothing but the lingering, the passing memory, of previous sense perceptions when the horse was still alive. The concept horse, on the other hand is a derivation, an induction acquired by the active intellect, by the immortal *nous poietikos*. The genus horse is immortal; and it is, so to say, God's horse. But the ordinary domestic horses are the mortal transient perception of the inferior *nous pathetikos*. God, concluded Gersonides, knows only the reality of the genera, and He does not meddle in the myriad, arbitrary sub-lunar special cases.

Gersonides' concept that God can only be approached through this heavenly faculty of the active intellect, is an insult to all the non-philosophers and simple folk. Only the sage who learned Carl von Linne's binomial system of scientific nomenclature by heart would be worthy and qualified to be accepted among the immortals in Heaven—and may all others be damned. That would be the ultimate consequence of Gersonides's outlook if we were to believe in it today.

Another well-known Jewish thinker in Spain was Hasdai Crescas (1340–1410) who was a welcome break away from the brainy and disgusting concept of the inhabitants of heaven. Crescas was perhaps more a rabbi than a philosopher; and, for example, he dared to believe in the actual infinite. Aristotle was, in Crescas's eyes, not a great thinker who sometimes failed, but a harmful nincompoop. Let us for a moment look into Aristotle's curios sophisms and those of his opponent, Crescas.

Aristotle once maintained that a wheel of infinite diameter cannot rotate. Modern physicists would agree…but for more lucid reasons that those adduced by Aristotle. He reasoned as follows: the distance between the spokes increases away from the hub. This distance would become actual infinite at an actually infinite distance from the hub, and an actual infinite distance cannot be transversed. Aristotle failed to notice that the velocity of the rotating movement would become infinite too, and this invalidates his argument. A modern thinker would adduce a different argument. There is an obvious limit to the velocity of the rotating spokes, increasing away from the hub. The velocity cannot surpass the velocity of light; and this means that the maximal diameter of the wheel is limited by its angular velocity. An angular velocity can never become actually infinitesimally small; hence its diameter cannot be boundless.

The arguments adduced by Crescas to counter Aristotle are, to put it mildly, weird. A spoke of infinite length has no end. We cannot indicate a point on a spoke where the distance from the hub passes from finite to infinite. So this distance is always finite, and infinity is out of the question. There is also no infinite distance between the spokes. How, from this obvious truth, must it follow that a wheel of infinite diameter must rotate befuddles my mind. But Crescas indeed committed this unbelievable blunder. The elementary mistake is that he ignored the difference between an actual and a potential infinite. The distance between his finger and the hub is a potential infinite that never becomes an actual infinite when he moves his finger away from the hub. It will always be a growing finite distance. But Aristotle had a different problem in mind. His starting point was a wheel of an imagined actual infinite diameter; and it is the imagination that lets us down.

Crescas was so utterly confused that he repeatedly committed the same mistake when he talked about the eternal past. He believed in its existence. He argued that infinite past simply means that prior to each time interval there exists a preceding interval ad infinitum. This reasoning also suffers from more than one error. First of all, Crescas confounded existence with imagination. He did not see that we may imagine things that do not exist. We may of course imagine how we would increase the length of, for example, a line by adding on sections of finite length. But the length would in reality never become infinite. We are again confronted with a potential infinite.

His second mistake was that he performed in his imagination an act in contradiction with reality. He added an infinite series of time intervals backwards in time, while, in reality, time flows only forwards.[4] But Crescas also had moments of lucidity. He understood the deficiency in Aristotle's definition of space. Space defined as that which lies outside the surface of objects must imply that there is no space without the presence of objects. Crescas revised this erroneous definition thus: "Space is extent existing independently of the presence of objects." Modern, theoretical, and relativistic physics teaches that a universe without objects would be boundless space, the more objects, the smaller the universe. This is the concluding statement in Albert Einstein's "The Meaning of Relativity."

How did Crescas react on emotional problems? Alas he was indeed deeply and tragically involved in the subject. Crescas, the rabbi, felt an intense sympathy for the fate of his congregation. His principal work called *Or Adonai*, which translated means "the Light of the Lord," tells about the light of his love, his only son. He experienced the harbinger of bad times. In 1391 his son was murdered by his persecutors.

Notes

1. See chapter 4.
2. See *More' Nevukhim—The Guide to the Perplexed* and his treatise on "The Art of Logic" 1:2.
3. See *Ethical Writings of Maimonides,* edited by R. L Weiss with C.H Butterworth.
4. See chapter 1.

7

Spinoza: The Aftermath
of the Spanish Intermezzo

The Spanish Intermezzo had an aftermath of monumental consequences in the world of philosophy: The system of Baruch de Spinoza. I know that Spinoza lived after the dawn of the Reformation and that the Reformation does not belong to the Middle Ages. But it is equally true that Spinoza's break away from the Middle Ages cannot be explained without going back to the Spanish past of Spinoza's family life and his Spanish-Jewish tradition.

When Baruch de Spinoza (1632–77) was born, the expulsion of the Jews from Spain was already history. Though the history of his family, of his father Michael and his mother Hanna, started with its new life in Amsterdam, his spiritual past—the philosophies of Maimonides, Gersonides, and Crescas, were still vivid reality. One should, however, bear in mind that Spinoza could not possibly use their philosophical systems as a point of departure. There simply was no such system.

The aforementioned three were too much at loggerheads about the problems where Aristotle had been right and where he had erred. There was, however, a common medieval mentality and a common way of thinking. This explains how it came to pass that Spinoza used the same point of departure. He just followed the old habit and opened his discussion with the essence of God. But Spinoza, on the other hand, was the first who challenged the authority of the Bible. This authority had to be replaced by the authority of human reason: the age of Descartes, Leibniz, and Newton had dawned.

Spinoza had to start with an invention of a new and reasonable definition of God. He said, "God is the absolutely infinite Being, a substance composed of an infinite number of attributes, each of them expressed as an eternal and infinite Being." These attributes have nothing in common with the attributes expounded in the notorious doctrine of Maimonides. For Spinoza they were not properties nor predicates, and he used Maimonides' term for something totally different. This was a very confusing habit and made it hard to follow him.

"Attributes" in Spinozistic terminology means "worlds." The physical world of space and matter is one attribute, just as the world of the mind is another. Spinoza believed that the number of God's attributes, or worlds, is infinite; but we mortals are able to perceive only two of them, the physical and the mental. All others are forever concealed from our perceptions

But the real breakthrough and fundamental novelty was the supposed relation of the attributes with God. They were no more regarded by Spinoza as independent worlds, each subject to its own system of causal laws, as Descartes and Leibniz believed: God—an intermediary Who prevents the worlds from falling apart. But on the contrary, Spinoza's attributes were all aspects of the same infinite Being, God, and they reflect the Substance God, every aspect in its own way.

This was the real slap in the face of the dualism advanced by Spinoza's friend, René Descartes. We may regard Spinoza, because of his new vision, as the pioneer, or, at least, the harbinger of the idealistic school that proceeded through Leibniz, Hegel, Fichte, and Fechner to Gerard Heymans. But how did Spinoza define the world of space and matter?

Let me take his tenet directly from the original Dutch text and quote it from the Appendix of the *Lexicon Spinozatrium*. Space, which he called *uytgebreidheid*, is indivisible because "divisibility of space would mean divisibility of God, which cannot be applied on Him.[1] And here I interpose my serious objections. To proclaim something a fact does not turn it into fact. Space is, according to our experience, divisible into finite volumes and it is measurable in units.

Spinoza's statement and definition of space have been driven to distraction by a contemporary of Isaac Newton, Henry More,[2] who entertained the mystic view that empty space, the cause of gravity, should be regarded as the Prime Cause or God Himself. Let us suppose that More is on an errand and searches for an empty bottle. Would he ask the dispensary to "give me a liter of our dear Lord?" I feel it is a blasphemy to identify God with something that is divisible. *Nihil,* the Zero or the Nothing that created the world is indivisible.

Let us have a look at Spinoza's definition of the other perceivable attribute, the mental world. The Dutch expression for mental activity in Spinoza's vocabulary is *"denking,"* and he said "the independent mental activity is infinitely perfect in its kind and an attribute of God."[3]

It was a fundamental mistake of Spinoza's that he regarded the world of space (*uytgebreidheid*) and the mental world (*zelfstandige denking*) as standing on the same level as the equivalent of God. On the other hand we have established that space is a sense perception—hence indirect information—in contrast with the mental world, thinking and feeling, which is a direct experience. Furthermore we reasoned that the chain of changes that take place in both the world of space (and matter) and in our mental world is unthinkable without the existence of time, which has been created. From all these considerations we had drawn the conclusions expounded earlier that the two worlds, the world of the mental reality and the world of space and matter, which derive from our sense perceptions, have been created out of nothing.

God cannot be identified with the world. Spinoza's slogan *Deus sive Natura*—which means God, in other words, nature—has to be rejected.

In Spinoza's pantheistic view, the world is eternal because God is eternal. Spinoza knelt for the same idol worshipped by Maimonides—Aristotle, who believed in an eternal world kept in motion by an eternal Prime Mover. We do not find any words of commendation in Spinoza's writings with regard to the CA as formulated by Saadyah Gaon. In this respect he did not follow Gersonides, who, as I have explained, did not mind to oppose Aristotle's views on time.

Spinoza's absolute submission to Aristotle was a fateful weakness and was one of the main factors that kept the school of the idealistic philosophers, who preached in favor of the *Creatio ex Nihilo*, apart for centuries.

But Spinoza was in one respect a true follower of both Maimonides and Gersonides. He followed the Iberian custom to search for God, but not out of scientific curiosity. He wanted, like Maimonides and Gersonides, to immortalize his soul. They all ignored that we cannot penetrate deeper into the unknown than the natural limitation of our mind and of our senses. Today we know that this was their lack of modesty.

The best we can do is to try and take stock of our human situation by straining our common sense and our moral sense to the utmost, to the very limits, of our limited abilities. The only reward is peace of mind and this is less than the unattainable object of Spinoza's eternal bliss in the great beyond. It is, however, of interest to explore the route proposed by Spinoza. Let us recall that the thinkers of antiquity discerned two steps of mental activity: The lower *nous pathetikos*, or passive intellect, and the *nous poietikos*, the higher active intellect. Maimonides and Gersonides just walked in the footsteps of these sages.

Spinoza, on the other hand, believed that our ascent to heaven has to be carried out in three steps of cogitation. He called the lowest level: *cognitio primi generis*, which is the knowledge directly derived from sense perception. This cannot be genuine knowledge, as the deductions are fallible and confused. We may regard this level as more or less the equivalent of Aristotle's *nous pathetikos*. Incongruous dreams, hallucinations, and wrong and garbled concepts are inevitable.

But every human being endowed with a normal amount of intellect is able to take the next step and reach the level of simple rational thought—the *congnitio secundi gerneris*. It is the level where scientifically adequate deductions are made, the starting point of genius reasoning on space and movements. But the highest level is the realm of God's own timeless thinking, where we have left the changes of the transient world behind us. Here reigns the truth. The divine truth is self-evident and knowable through our innate standard of knowing the correct from the false. God is the great and perfect computer, not an inanimate machine, but the Being of pure timeless mathematical logic. The endeavors of the human mind to rise above the *cognitio secundi generis* are meant to turn man himself into such a computer, though of a more modest size, thinking the thoughts of God. In this highest realm man and his soul have freed themselves from the transient sub-lunar world receiving the *beatitudo,* the eternal bliss of immortality.

Let us compare Gersonides's highest aim with this ideal of Spinoza's translated in modern terms. Gersonides' intention would mean that the teachers in systematic biology, thinking in genera, are the only real immortals. In Spinoza's eyes this highest good was obviously reserved for the smart teachers in higher mathematics. May the half-wits and flunkies be damned.

Spinoza's study object, the timeless logic as the highest reality, reminds us, in a way, of Albert Einstein's four-dimensional space-time, where time is merged through its multiplication by the velocity of light into a fourth linear dimension. Einstein's world seems in a certain sense to be a timeless one in analogy with Spinoza's timeless ideal reached through "*scientia intuitiva,*" a kind of intuitive knowledge of the highest grade of absolute truth. Spinoza might have been delighted by Einstein's theoretical discovery. But this would be a fallacious reading of Spinoza's intentions. Einstein was not a Spinoza. His writings were not a guide to the salvation of the soul, but a reasoned description of the world of our senses, which takes account of all the modern discoveries. His four-dimensional space is not God, but God's creation as evidenced by the corrections of Einstein's system by his colleagues Alexander Friedmann and Willem de Sitter.

This means that Spinoza's picture of the world cannot be applied on the modern views. His main error was that he confused acquiring knowledge about a subject with becoming part of a subject through knowledge, which is an absurd idea. We cannot change our fate as mortals by acquiring whatever wisdom in this respect. We are all in the same boat, the wise with the half-witted. I know by experience that it is a strenuous effort to think intensely about all the consequences of Einstein's four-dimensional universe. It occasionally may cause me a headache; but I do not feel that the exhilarative experience of a new insight would enhance the immortality of my soul.

From where did Spinoza receive his absurd ideas? I believe from the superstitions of the old Greek philosophers, who did not know yet that there was a connection between brain and thought. To them, thought came from heaven from the *nous poietikos*.

That Spinoza has guided us over the brink and enticed us to fall from grace is much worse, it is really sinful. He encountered on his way to heaven, in his three-staged spaceship of cogitation, an unexpected hurdle—his sense. It was, for Spinoza, a much more serious stumbling block than for Maimonides. Maimonides committed the sin, debasing the Torah by regarding it as a means to attain something imaginary that he valued as higher than moral perfection. But Spinoza descended still deeper into hell. He first debased the Torah as if it would be just a piece of *ius voluntarium* not better than any other civil code. And then he debased even the civil code; to him the law was just a means to bridle the dumb. The wise and the sage do not need it, because to Spinoza sensible people didn't sin.

Moral sense was to Spinoza not so much a sense to be heeded, but a prejudice that a wise man has to overcome through reason. Moral sense obliges us to submerge our interests in the higher ideal, the understanding of the whole order of nature. Spinoza believed that a free and wise man should feel morally neutral with regard to the interests of others. This was a sinful attitude. And I can sum up my feelings by quoting one of Dante's rare appealing statements from his *Inferno*: "The deepest places in hell are reserved for those who stay neutral where moral issues are at stake."

Spinoza calls our drives "passions" that have to be kept under control. This is a most confusing term. Especially when the word passions is used in the context of morals. Passions may mean sexual passions, which are not necessarily sinful. One of my friends, a very serious Prot-estant priest, conceded that he got his sexual passions in order to have them tickled. Let us add that our passions, including sexual passions, are as much subject to Heymans's principle of objectivity as all the other drives. One is not allowed to satisfy one's own passion at the expense of the feelings of others.

But the most serious question we should ask ourselves is this: does the urge to perform good works not also belong to our passions? Indeed it does; and it was Spinoza's intention that we should confine this urge as well. He stultified his own moral sense, because he re-garded it an impediment that hindered him to rise to *scientia inutuitiva*. This attitude reminds us, firstly, of Immanuel Kant who tried to cre-ate a difference between passion and duty, unaware that the urge to perform a duty belongs to our passions, the good ones. And secondly it reminds us of a much worse example: Joel Goldschmith, who preached that moral virtue is an impediment in the search for mysti-cal union with God.

Why do I regard Spinoza's endeavors as wicked seduction? Because civil codes are not meant to bridle the dumb, but the bad characters who misuse their intellect. Criminal jurisprudence serves to sober their evil intentions. If we have to believe Spinoza's contemporary (he must have lived in a kind of moral catalepsy), his friend J. Colerus, a Lutheran clergyman, told that he studied small gnats under the newly invented microscope and that he caught spiders and observed, with a kind of detached pleasure, how flies fought for their lives in the webs. If the souls of the dumb have to forgo postmortal survival, Spinoza must have maintained that the far dumber animals have no soul at all.

Though Spinoza derived his philosophy from two Jews, Maimonides and Gersonides, it does not belong to Judaism anymore. It is general philosophy—though we may still call it a religious philosophy. Spinoza emancipated himself, or better, became estranged from his Jewish past. He stealthily scaled the walls around the spiritual ghetto of Amsterdam and inhaled the air in the gentile world. Those inside the Jewish world did not pay much attention to him because they were too deeply study-ing Talmud. But then Spinoza committed something unheard of. He effected a breach in the walls from the outside and left the rabble stand-ing in the draught. I am aiming at the disrespectful way of his ridicule, which he brought on everything connected with religion, especially with Jewish religion. In his aloofness he could not foresee that a few centuries later everyone who uses his reason according to Spinoza's prescript would understand that the view of the rabble prevails and not the philosophy of Spinoza—that God created the world and that the most valuable gift the Lord presented to man was not reason, but moral virtue.

The first faltering steps of the early freethinkers were fraught with the evil consequences of their missteps. Spinoza did not escape this fate, but even a misstep may be a blessing in disguise. It became more and more of a necessity to free oneself once and forever from the oppressive, medieval authority of the Bible.

Spinoza broke radically with religion. This break empowered the thinkers after him to reappraise the Bible from a distance and from their new vantage point. They became aware that the old fundamentalist's standpoint, the absolute faith in the letter of the Bible, was wrong. The Torah and the Old Testament are not what they say; but what they convey—the message.

Notes

1. *Wij uytgebreidheid een eigenschap van God stellen te zijn 't welk aan God algeheel niet kan toegepast worden.*
2. See Moshe Jammer's *Concept of Space.*
3. *De Zelfstandige denking is oneyndig volmaakt in zijn geslagt en een eigenschap van God.*

8

The Very Few Jewish Idealists after Spinoza

Spinoza could never have made an impact on Christian philosophers if he had lived in the time of Maimonides. His voice would have been muffled by the almighty Catholic Church. The Reformation suddenly encouraged some free thinking. The gates were set ajar and the trickle turned into a flood. Spinozistic idealism became the fashion among the Christian thinkers. But the Jews shut themselves out from this new mode, not so much because of a reactionary state of mind, but because of much more lucid counterarguments.

But a handful of Jewish thinkers did not regard the arguments as watertight, and they were themselves previous to idealistic ideas. Let us take four examples, Nachman Krochmal (1785–1840); Solomon L. Steinheim (1789–1866); Solomon Formstecher (1808–1889); and Samuel Hirsch (1815–1889). Their proclivity brought them into some trouble and embarrassment. True idealism from Spinoza to Heymans is based on a dependence of the world of our senses, on the world of our mind (the modern view) or on the interdependence of the two worlds on God (the Spinozistic view). *Deus sive Natura*, Spinoza's slogan meant that God and nature are the same. Because God is the Eternal Uncreated, nature must be equally eternal and uncreated. This is a principle that can never be reconciled with the central Jewish precept to believe that God created the world.

I explained in the first chapters of part 1 that it looks, from a scientific point of view, even much worse for the Jewish idealists. God created the world. This is not just some Jewish and Christian article of faith. It is a scientific fact based both on reason and on observed facts.

How could the few idealistic Jews extricate themselves from this quandary? Because the principle of orthodox Jewry that the existence of the world depends on God, but that God does not depend on the world, is binding. They had to modify the imminence of idealism to become to a certain degree, eminence, a half measure that could not convince the orthodox.

Solomon Steinheim went about it the wrong way from the beginning to the bitter end. He started with an obvious fallacy: "*Ex nihilo nihil fit.*" I thought it impossible that matter has a temporal beginning. This conclusion would be a slap in the face for Judaism; hence he felt obliged to accept Creation as a doctrine. This meant that Steinheim had been influenced by a very silly philosopher, a certain Samuel Clarke, who dared to write the following nonsense:

> For since 'something now is,' it is evident that something always was, otherwise the things that now are, must have been produced out of nothing, absolutely nothing and without cause: which is a plain contradiction in terms. For to say that a thing is produced, and yet there is no cause at all of that production, is to say that something is effected, when it is effected by nothing, that is at the same time not effected at all.[1]

It is clear that the simple proof that time has been created knocks the spokes out of the wheel of Clarke's argument. Let us revise Clarke's sentence as follows: "For since 'something now is,' it is evident that something always was since the moment time existed."

If we want to appreciate the system of Solomon Formstecher, we have to realize that in his time, in the early years of the nineteenth century, a reasoned education did not rise yet to the present-day standards of exactitude.

Deductions bore often the features of exposition of intuitive, associative thoughts. This characterizes also the logic of Formstecher, who borrowed the foundation of his system from an arch-romanticist of his time, the idealist G.T. Fechner. If we overlook these imperfections we have to admire Formstecher's keen-witted intuition.

His point of departure was the system of Spinoza, but he deviated from his philosophy with the introduction of two new principles that were incompatible with the ideas of Spinoza. Formstecher understood that Spinoza's slogan, *Deus sive Natura* must be wrong. The Creator is not dependent upon the world; but the world is dependent upon its Creator.

Formstecher furthermore objected against another fundamental principle of Spinoza, who regarded the world of space and matter and the mental world as two equivalent attributes of God. Formstecher's point of departure was that nature, which we call the world of space and matter, is subordinated to spirit, which we call the mental world.

These are two essential strides in the right direction; an enormous progress compared with Spinoza's original ideas. When Formstecher turned to the subject of religion we have once again to admire the qualities of his intuition. He was like a man who first closes his eyes and then hits the nail with an infallible feeling: "God is the essence of ethics."

It is a matter of course that I cannot agree with Formstecher's final product. He calls the physical and the mental world his "nature" and his "spirit," two manifestations[2] of God. Though Formstecher is aware that they cannot be equivalent manifestations, he could not grasp yet in his time that the physical worlds, i.e., the worlds of space and matter, i.e., nature, is nothing but a mental structure, an interpretation by the human mind from sense perceptions that are data received from an external world—a world that turns out, upon closer examination, to be of a mental, i.e., spiritual, character.

If we are allowed to use Formstecher's term "manifestation of God" for what we usually call "creation of God," we should not regard the worlds of space and matter as a separate manifestation. On the contrary, the world of space and matter is a manifestation of God of the mental world, which is the manifestation of God, i.e., the creation by God. The world of space and matter is the mental world seen though the filter of our senses.

Another point in Formstecher's work, *Die Religion des Geistes*, is the sloppy use of the term "spirit." The spirit of God, the spirit of man, used in the same spirit; God, who is spirit, manifests Himself as the World-spirit, a spirit that is conscious of itself and of the world, its acme being the free consciousness of man's spirit.

This "system" is for me, at least, hardly comprehensible. It is confused mumbo-jumbo. Let me straighten out some of the most obvious misunderstandings. The trouble with the term spirit is that we mortals are unable to separate it from thinking and the catenation of thoughts. In time, thinking is unthinkable without time. Our innate habit to associate spirit with thinking implies, therefore, that talking about God as being Timeless Spirit is talking nonsense. The timeless cannot be grasped by the human mind. Timeless is tantamount to nothing. In short, we should not upset the term "spirit" when talking about God, and the term "Holy Spirit" as a translation from the Hebrew *Shekhina* is confusing.

The term "manifestation of God" is not to my liking either. I still prefer the term "creation by God." The term "manifestation" calls Plotinmus and neo-Platonic emanation to mind, and, as we have seen, it is a principle that is a source of trouble to understand Creation, because it suggests an interdependence between God and the world. That Christians have assimilated this principle in their faith is their problem.

The statement that God is the essence of ethics is objectionable, because of the consequence that God, our creator, would also be our Supreme Judge. We have associated God with ethics in a different way. God has instilled in us our moral sense as a faculty to judge among ourselves. We have to confess once again that common sense was not involved when we drew this conclusion; it grew out of an emotion, a feeling that moral virtue is a divine gift. We often hear the statement that humanism is the view of life that put man in the limelight and that religion is the view that puts God in the limelight. What I am putting forward is a third conception. God put man in the limelight of His creation, by endowing him with moral sense.

Formstecher believes that the message of Judaism is its mission to free Christianity from its pagan elements; and that Judaism has to prepare itself for that great day by discarding its particularistic ceremonial rites and laws. This was, in his time, the ideal of reform Jewry. Hardly an ideal, as we shall see in this chapter, but the disguise of a baseless shame of being different and, above all, fear of attracting attention.

But I have a better argument against Formstecher's view than just blaming him for a disused, insipid cringing from fear. I am convinced too that many religions, perhaps all the religions, have to rid themselves from objectionable articles of faith and habits, but it is strictly forbidden that they should be shut off from their past. Forgetting one's past means losing one's future. I confess, however, that I often feel a bit peevish when somebody suggests the word "tradition." Tradition may mean the product of the selective memory. The more unpleasant episodes of the past are forgotten just to make the past a more pleasant place in which our minds may roam. A sincere religion is an honest religion, aware of its mistakes of the past, which teach us how to make the future—and not the past—comfortable to contemplate.

The book by the fourth idealistic Jewish philosopher, Samuel Hirsch, *Die Religions-philosphie der Juden* expresses exactly my objections against Formstecher's proposal to water down Jewish religion to become merged in a tepid world religion together with an emasculated Christianity, which has attoned for its sins. Hirsh is of the opinion that Judaism should always honor its particularistic character, national and religious. He was a typical religious Zionist, though this special brand of Zionism predated Herzl's by several decades.

And Christianity? Christianity should first rid itself from its Pauline blemish. In the day of the Messiah, Christianity will become identical with Judaism. Indeed let us pray that the Messiah will appear soon, but I am not so optimistic. All these pious daydreams of Hirch's meet with my wholehearted approval. But not so the sort of philosophy and reasoning. Hirsch guided his train of thoughts along the tracks of Georg Wilhelm Friedrich Hegel (1770–1831).[3] Hegel was a man much less to my liking than Friedrich Joseph Schellingh (1775–1874) who inspired Formstecher.

Hegel developed an odd picture of the world in which a thesis encounters its opposite, the antitheses, and the inevitable contest engenders a higher synthesis. Hegel suffered from an obsession, the "*idée fixe*," that everything in this sub-lunar world is engaged in this dual— a combat between thesis and antithesis. But Samuel Hirsch believed in this though he modified the outcome of the eternal battles. A fight between a thesis and its opposite can never engender a higher truth, but the antithesis stabilizing the truth must annihilate the thesis symbolizing the lie. For example: Judaism symbolizing the truth must annihilate paganism, which symbolizes the lie.[4]

The last one of the Jewish idealistic thinkers of this period who deserves to be reviewed is Nahman Krochmal. He was the most popular among them, probably because he was the most Jewish. He wrote, for example, in Hebrew, and he was the founder of the "*Wissenschaft des Judendtums.*" We should not confound his *Wissenschaft* with adequate religious philosophical research. His *Wissenschaft*, or science, was partial—not to say biased. He only dealt with the historical aspects of Judaism. Solomon Schechter (*Studies in Judaism*) gave us a beautiful biography of Krochmal, full of unknown details, but all those vicissitudes do not bring us a step closer to his personality as an original thinker—only his work, *The Perplexities of the Time*, does that.

Krochmal described the history of the Jewish people as the development of its relation with the absolute. But Krochmal was a very complicated personality. Occasionally he even scares me. He was a typical eccentric who chose from a great variety of philosophical systems, whatever suited him, and he used these elements as building blocks for his system. The result is a not very consistent whole, not free from plain incompatibilities. He was, for example, a great admirer of Abraham ben Meir ibn Ezra (1092–1167) one of the very rare neo-Platonic thinkers who proclaimed that God is one. He said that God "made all, and He is all; He is all and from Him cometh all; He is the One, and there is no being, but by cleaving to Him."[5] Spinoza was certainly influenced by Ibn Ezra; hence his slogan *Deus sive Natura*.

On the other hand, Krochmal agreed that the only attribute that we may ascribe to God is His Nothingness. That he borrows this truth from the murky fantasies of the Kabalists and not from the crystal clear deductions of Saadyah Gaon does not impair the final conclusions concerning the goal of Krochmal's religious ideal—the total communion of the human spirit with the divine to be reached through the knowledge of God.

To be aware of God's Nothingness and nevertheless to strive after the communion of the human spirit with this Nothingness scares me out of my wits. Krochmal is advertising a classical, mystical aim, nothing better than the suicide of the mind. I am furthermore completely at a loss how we may glean knowledge about Nothingness, through even the greatest intellectual effort. I cannot agree with Krochmal's contention that the intense adoration of the Divine Nothingness may assure me of my everlasting life. It is a superstition that Krochmal borrowed from Maimonides and Gersonides—and adopted by Spinoza in a more brainy variation.

Krochmal's visionary ideas and fantasies concerning the development of "the nation" are not less disturbing. I still have to be convinced of consciousness transcending the individual. Even the feeling of some telepathic understandings with my nearest comrade (my wife) is a rare experience. Krochmal believed in the spiritual entity of a nation as a real, individual, being that in its development passes through the stages of birth, growth, and decadence. All members of the nation participate in this "spirituality" of the whole. The development of man as a spiritual individual is only an intermediate stage, the acme and apogee is the "spirit of the nation." The Jewish people and nation have been able to survive the natural demise, which is the fate of every nation, through its faith.

I do not deny that the members of a nation are bound together through common experience. They are the heirs of a common past. But this feeling of fellowship and solidarity does not turn fellowship and solidarity into an individual being with its own mind and consciousness. It is an idea that reminds me of some aspects of Gustav Jung's "collective unconscious," in which the spirit of the individual is but a building-stone of the spirit of the entire society, heir of its entire past—the prehistoric past counted in. It was a daydream that began to haunt Jung's mind; but then he was an incurable mystic anyhow.

Krochmal's mistake in endowing the rather abstract nation with a concrete individual "soul" becomes understandable if we try to grasp his conception about science and its task. He understood quite well that the purpose of science is to discover the principles and general laws that guide the phenomena of nature. He maintained, however, that these laws were of a most universal validity; and Krochmal magnified the universality of their validity in his mind to a degree that these laws become a concrete individual being with an eternal life. The laws were detached from the phenomena guided by them and began to lead their own lives. It is a beautiful example of sophistic deception of the senses.

After my exposition of the various standpoints of this foursome, the four principle idealist Jewish thinkers at the beginning of the nineteenth century, you will certainly ask me a justifiable question. Why did I push them straightaway to the foreground, into the limelight, before I even mentioned the names of the great multitude of thinkers who appeared on the stage after the Reformation? I have several answers. First of all, the four were religious philosophers. Secondly, they were certainly important and deserve a prominent place, though we do not agree with their principles. But lastly, though not least, they were four odd men, not in the mainstream of Judaism, who may easily be forgotten and I therefore wanted to untie the knot in my handkerchief as soon as possible.

Notes

1. This sophism, dated 1704, was reprinted in *The Cosmological Argument*, by William Row.
2. "Manifestation" meaning more or less the same as "attribute" in the vocabulary of Spinoza.
3. Hegel was a German philosopher whose views greatly influenced subsequent philosophy. He wrote on logic, ethics, history, religion, and aesthetics.
4. All this is antiquated nonsense and only of historic interest. However, I recommended to read the chapter "Post-Kantian Idealism in Jewish Philosophy," in Julius Gutmann's *Philosophies of Judaism* with care. He handles the subject of Jewish idealism in the nineteenth century with admirable objectivity and he does not squabble with the various thinkers about the inconsistencies like I do. See bibliography.
5. See Julius Guttman, *Philosophies of Judaism: A History of the Jewish Philosophy from Biblical Times to Franz Rosenzweig.*

9

The Legacy of Spinoza

The medieval Catholic Church and its ostentatious display of powerful executive authority of the Christian sinful faith, bore its own chemistry of internal fermentation and decay, though the Reformation did not manage to bring it definitely to its knees.

These events did not take place in Germany, Switzerland, and France; not within the spheres of action of Luther, Zwingli, and Calvin, but in the Netherlands and in England. Spinoza happened to live in Holland, and that was a fortunate coincidence because of the invention of the telescope and microscope in that same country a few decades before him.

Lenses were already known in antiquity. They have been dug up by archeologists in Nineve, and were in practical use during the Middle Ages, to correct impaired eyesight. But, oddly enough, it took centuries before somebody discovered that a few lenses properly arranged and encased in a tube may serve as a mighty tool to examine the macro- and microcosmos. The inventors of the instruments were Dutchmen. The telescope was presented by Lippershey on 2 October 1608,[1] and the microscope was presented a few years earlier by Hans and Zacharias Janssen in 1590. Spinoza was not only acquainted with these new objects, he was himself a professional producer of lenses and he was a most eager researcher.

The intellectual world suddenly stopped consulting Aristotle's works. If they wanted information about phenomena in nature they began to rely on one's own senses and intellect. The seventeenth century was indeed the dawn of modern science. The birth of free research went hand in hand with the birth of free thinking. The new ways cleared by Spinoza were duly appreciated—especially in Germany.

The German philosophers were broad-minded enough to realize that Spinoza was human and fallible. A great many began to tinker with his system; and a few began to think seriously about its shortcomings. The great many put the system to shame, but a few cleaned it from its obsolete neo-Platonic and Aristotlelian characteristics. The fallacious principle that the physical and mental world are equivalent "Attributes" of God went by the board, and one began to understand that the world of the sense should be regarded as the shadow of mental reality.

The first great German to adopt Spinoza's system was Gottfried Wilhelm Leibniz (1640–1716). René Descartes could not convince him that the Good Lord should be bothered through eternity with the task to uphold the connection between the physical and the mental world. Spinoza's friend Arnold Guelinck compared the physical and the mental world with two clocks well synchronized. God holds His one hand on the hand of one clock, and His other hand on the other clocks; and in this way He fulfills the tedious rut to move them at the same rate. This symbolizes Descartes's concept.

Leibniz wanted to release God from this boredom. Why not suppose that God consulted His own clock at the great event of creation, and set the mental and the physical clock accordingly. They are of such a divine precision that nothing incongruous will ever happen. Geulinck's illustration was an excellent, popular description for the great public. Leibniz rejected Descartes's concept, which he called the *"influx physicus,"* and replaced it with his own principle, the *"harmonia praestabilitata."*

A few centuries elapsed and Gerard Heymans brought Geulinck's illustration to a greater perfection. Spinoza had only one real clock in mind, but it is shielded from the human eye—this is God's own clock. This divine timepiece is mirrored by two looking-glasses creating the image of God's two known attributes, the physical and the mental world. This modification releases God's hands from their task. The reflected images of a clock are always synchronized without the clock itself, and this does not ask for further explanation.

The modification proposed by Heymans was another improvement on the model. An image is dependent on the original that created the image, though the existence of the original does not depend at all on the existence of its image. But this was not what Spinoza had in mind. He believed that the two are interdependent (his slogan again). God and nature are, in Spinoza's philosophy, coexistent—the one is unthinkable without the other.

It was this obvious mistake of Spinoza's that had to be corrected, and this correction has been executed by the philosophers who came after him. The last one being Gerard Heymans, who did away with one of the mirrors. The real clock symbolizes the mental worlds, and this image in the mirror symbolizes its reflection through our senses, the physical world. The mental reality exists always as long as we live in our mental activities, but the physical world depends on our sense perception. We close our eyes and block our ears, and we shut the physical world out. Many a modern idealist identifies the real, mental clock with God, denoted as the Absolute. We have seen that this view is erroneous. The mental world has been created, out of Nothing.

Leibniz unified the physical world and the mental by atomizing the cosmos into "monads." He called these smallest particles explicitly "seelen" or "souls." The world we perceive as physical is thus composed of an infinite amount of little souls—"monads"—who are unaware of each other. In other words, "monads have no windows."

I feel once again obliged to interpose a remark. This time a serious warning. The theory of Leibniz and those of the idealists after him, should not be confounded with the erroneous identity theory of U.T. Place and his associates of the twentieth century. They maintained that a thought consists of numerous small brain processes just as a cloud consists of numerous droplets. The atomized brain processes are, in Place's eyes, physical reality. This materialistic view is exactly the opposite of Leibniz, who realized that his "monads" must be mental reality. Leibniz's view is idealistic, and he based his theory on intuition—it was not a reasoned deduction. But that did not alter the fact that Immanuel Kant and even Bertrand Russell were duly impressed.

The idealistic worldview of Leibniz was rounded out by his contemporary, Bishop George Berkeley (1684–1753). Critics of his time accused him of throwing the philosophical world into confusion. Their remarks were not baseless. Berkeley was more than a bit confused himself. But we should not forget that Berkeley seriously reflected upon the essence of space and matter, that he laid the foundations of the empirical way of thought, which means that he understood the importance of introspection, essential to acquire knowledge about our own mind.

Berkeley is known on account of his slogan "*esse est percipi*," which is generally translated as "only the perceivable exists." But Berkeley's real intention was that the existence of anything not being observed at a certain moment should be denied. His first question was "what engenders perception?" His answer betrayed him. "Both perception and to be perceived depend on a divine will. A sense perception exists by the grace of the Holy Ghost. It must be clear that the world of space and matter is nothing but an appearance of a world of moral reality." Though we may regard this statement hardly as an outcome of scientific reasoning, we have to bear in mind that idealism with all its weaknesses became a mighty bulwark through Berkeley's influence, which stemmed the tide of its opposite "monism," materialism—which teaches that matter is the only reality, while mind and soul are dismissed as hallucinations, an "epiphenomenon" of no consequence.

Berkeley's fertile mind opened a flourishing period of idealism in his own country—England. I mention the various thinkers only by name: Edward Caird; T.H. Green; F.H. Bradley; Andrew Seth; Hastings Rashdall; J.E. McTaggart; and finally T.L.S. Sprigge who wrote *The Vindication of Absolute Idealism*. We shall get better acquainted with only one of them. What were the views on space and matter in Berkeley's time? There were quite a few scientists who took Descartes's duelist views for granted. They even refined his erroneous system. I refer to Peter Cassendi (1592–1655), Robert Boyle (1627–1691), and his pupil John Locke (1632–1704). In Isaac Newton's (1642–1727) system the old dualism was even the mainstay.

But we have to dig deeper still. Where did these seventeenth-century pictures of the world come from? Mainly from the old Greek systems of Democritus and Epicurus. The world is composed of numerous insensible corpuscles or atoms only endowed with so-called primary properties, which are shape, size, bulk, and motion. All the other attributes are regarded as so many secondary illusions, evoked by our senses, or as Democritus declared, in the fifth century B.C.

Hot and cold are appearances, sweet and bitter are appearances, color is an appearance; and in reality there are only atoms and the void. This sort of distinction between primary and secondary attributes seems odd and arbitrary in the eyes of a modern scientist. Shape, weight, size, and motion are as much a making by our sense as color, temperature, and taste. It seems to me however that the old Greek had some vague notions about a distinction we today call the difference between the physical world and the world of the physicist.

The physicist intellectualizes our everyday gestalt world when he tries to explain our sense perceptions. He rebuilds our gestalt world into a causalistic comprehensible system. This system is an abstraction in which a part of our various impressions form our sense perceptions—though not all of them go by the board, such as color, taste, and smell. Color and sound are translated into the frequencies of vibrations; the sensations of hot and cold are translated into velocities of the atoms and molecules; those of taste and smell are translated into properties of liquids and gasses. Physicists reduce the world of our senses to energy, time, and space; but only a few are aware that this ultimate picture does not take into account that the source from which it is derived is nothing but the interpretation of sense perceptions. They objectify this interpretation as if it reflects the true external world. In short, the world of the physicist is essentially a materialistic one.

The dualistic character of Robert Boyle's world speaks from his own words: "The corpuscles of Democritus strike on the sense organs and cause motions which are transmitted to the brain," so he says. John Locke maintained the same: "The mechanical impact of the insensible corpuscles is transmitted by the nerves to the brain generating mental sensations called ideas." This dualistic concept is saddled with the unsolvable riddle of how mechanical motions may ever become mental emotions.

The startling point of modern Idealism is that the world, as we perceive it, is not identical with the external world. The external world may only be of the same kind as that which makes perception possible, and this is mind.

But George Berkeley gave a different answer. He became so frustrated by the paradox how corpuscles made of matter could ever cause mental impressions, that he rejected the objective existence of matter altogether. John Stuart (1806–1873) wrote, more than a century later, that objects are entities bearing the permanent potential of being perceived. Berkeley, on the other hand, negated even the permanence of their existence. Objects only exist as long as they are perceived. The permanence of this world was only guaranteed by being permanently perceived though God's eye.

An important Berkeley biographer J.O. Urmson (*Berkeley*) commented that this skepticism entails a very obvious difficulty that Berkeley never solved; that the perception of the ever-changing world through the divine eye evokes ever-changing sensations in God's mind. And if so, how could this tally with the current conception of an eternal immutable, God?

Typical idealists who are not committed to believe in an immutable God are free to believe in Spinoza's slogan that God and nature are one and the same. But pious Christians, like George Berkeley the Bishop, were in a quandary. Berkeley was in need of a perceiving eye to maintain the existence of the world—a mortal eye could not do. One of the fundamental principles, from a biblical as well as from a scientific point of view, is that perceiving creatures came into being after the creation of the inanimate world. This is hardly feasible if the existence of the inanimate world would depend on its being perceived by an animate being.

All these blatant inconsistencies in Berkeley's system show that he cannot be regarded as having been a true idealist. He grappled with the problems, but he certainly was, with Leibniz, one of the first heirs of Spinoza's legacy.

Note

1. The Italian Galileo was the first who pointed the new invention in the direction of the stars.

10

The Development of the
Concept of Space-Time

The seventeenth century has been witness to the new developments of the views on space and time. The Aristotelian definition of the void—that is, space—was that which surrounds objects with the consequence that without the presence of objects the void would not exist.

It was high time to dispose of this paradox, and Isaac Newton reformulated space. He said that "absolute space in its own nature, without relation to anything external, remains always similar and immovable." But he rejected forthwith the consequences of an absolute void. That gravity should be innate, inherent, and essential to matter, so that one body may act upon another at a distance through a vacuum, without the mediation of anything else, is to me so great an absurdity that I believe no man who has, in philosophical matters, a competent faculty of thinking can ever fall into it. In other words, space is not a void, not a vacuum, it is filled with something unobservable that transmits the action of gravity.

But the original postulate that space is an active conveyer of action is very old and against the idea of Democritus. "Ether-whirls" twirl the heavenly bodies around their axes and blow them in circles through the cosmos. It was a most inspiring idea that fired the imagination of the scientist in the seventeenth century. "Ether" became to Newton the transmitter of the action of gravity from one body to another at a great distance. The cosmos and its space looked, before Newton's work, like a huge three-dimensional map of ether-wind currents that carried the stars around on the wings of huge hurricanes.

This was an idea that inspired Descartes as well as Leibniz, but their views of a windswept universe were discredited even in their very lifetimes by the sobering discoveries of Newton. The velocities of the rotation and orbits of all the celestial bodies may be predicted with great accuracy by applying on them the gravity equations of Newton. These laws predicted that planets may orbit in planes at an arbitrary angle with the rotational axis of the larger body. If a planet orbits in the plane that coincides with the equator of the larger body, it may go round in the direction of the rotation of the larger body or in the opposite direction. In other words, the orbiting of planets defies the laws of aerodynamics, which are supposed to rule the ether-currents. We may as well conclude that ether-currents do not exist.

But this forgone conclusion did not yet kill the ether theory. It stayed alive for a long time after, albeit in a revised form. Ether was the postulated filler of space and the carrier of light, the so-called luminiferous ether composed of extremely fine particles of subatomic dimensions. Postulating a concrete, but in principle invisible, medium as a last resort to explain phenomena is regarded by every serious scientist as an awkward stop.

Is there anything real behind ether? It seemed to be the case after the theory of Dutch scientist Christiaan Huygens (1629–1695) prevailed over that of Newton's. Huygens regarded light as vibrations propagated through the ether, and Newton believed that a light-ray is a salvo of very small corpuscles. Huygens's theory seemed to be much more attractive after linguist cum physician and physicist Thomas Young (1773–1829) discovered the phenomenon of light interference, and even more so after the English chemist Michael Faraday (1791–1867) discovered that the phenomenon of electricity and magnetism are two inseparable twins, and that an electric current evokes magnetism, and a moving magnet creates electric current. Then it turned out that light consists of a vibrating electric and magnetic phenomenon. The one vibrating in a place at a right angle to the other. Or, in other terms, the propagation of light follows the laws of the electric magnetic field theory of James Clerk Maxwell.

Should light consist of vibration then we should expect that "something" vibrates. To maintain that the void, the nothing, vibrates would mean that there is nothing that vibrates; that the vibrations are, in other words, absent. Would this absurd conclusion not imply that there must be something we call ether? That space contains something and that space is not an empty void? But what is the role of this something in the equations of Maxwell? It turned out to be one of the great modern paradoxes—the equations "burn the Thames." They do not imply the absence of a medium at all; on the contrary, they seem to imply the absence of a medium that the nineteenth century called the "luminiferous ether."

It was, first of all, Maxwell—not Albert Einstein as generally supposed—who dismantled the redundant scaffolding of the ether that surrounded his splendid edifice of theoretical physics. On the other hand, Einstein was immediately prepared to accept all the consequences of the paradox that light is not transmitted through a stationary or moving medium. Einstein banished the ether out of science, and took this step in the defiance of his friend, Dutch physicist Hendrik Anton Lorentz (1853–1928), who tried to get around the ultimate consequences of the constant velocity of light by introducing the postulate that bodies at very high speed are flattened against the resistant ether, which he supposed to be fixed in space as solidly as a wall. M. Jammer, who wrote in 1954 the book *Concepts of Space*, said that

> epistemologically, Lorentz's theory shows its unsatisfactory character by the fact that it ascribes to the ether or absolute space certain definite effects, which by their very assumed existence preclude any possible observation of the ether. Similarly, all other experiments to identify the ether as a privileged system of reference had to go by the board.

Ether, therefore, is a nonexistent substance, though space is not "nothing" because it has been created out of Nothing. Furthermore we have to realize that space cannot be of an actual infinite extension (Einstein's equations take proper account of this) nor could it be possible to subdivide space into actually infinitesimal small volumes. Here, Einstein's equations could not take account of this law. His space is a "continuum."

Modern scientists are hard at work trying to modify Einstein's theory, or at least to supplement it with the introduction of the principle that space becomes foamy and its behavior unpredictable below the scale of subatomic dimensions. Roger Penrose of Oxford proposed that the breakdown of conventional geometry may already be felt at about the diameter of an atomic nucleus. (I am referring to his "twistor" model of space, his project of theoretical research in the possibility that space consists of subatomic space-particles.)

The latest fashion in theoretical physics is the hypothesis of a ten-dimensional space-time, which means that nine dimensions are spatial and that the tenth includes the time parameter. This contraption of a ten-dimensional cosmos is supposed to be inhabited by one-dimensional "super strings" of energy just 10^{-33} cm. long. Author Gary Taubes wrote in *Discover* magazine, November 1986, under the article "Everything's Now Tied to Strings" (endorsed by the American Association for the Advancement of Science) that "to get experimental predictions out of superstrings, someone must figure out how to collapse the ten dimensions of the theory into the four dimensions of reality."

I have serious misgivings concerning the logic of this comment. First of all, I dislike the juxtaposition of the four dimensions of reality and the other six dimensions. Are these six supposed to be less real than the other four? Secondly, I ask something about the reality of the Einsteinian four dimensions of space-time. Reality here does certainly not mean realizable in our imagination. We are not able to imagine four spatial dimensions; we are conditioned to imagine only three Euclidean dimensions of space—and time is something else. We have viewed this space-time from a broader, I may call it metaphysical, point of view. We derive Euclidean space from the perceptions through our space-sense, which means that even the three-dimensional space of our everyday experience is not a reality but a mental construction. The only parameter that seems real to us, which means a parameter in the mental reality, is, as we have discovered, time.

But even viewed from a narrower point of view, regarding Taubes's comment only within the realm of natural science, I hardly feel a need—from a purely philosophical point of view—to collapse the ten dimensions into four. Let us once again quote Einstein: "Science is the attempt to make the chaotic diversity of our sense-experience correspond to a logically uniform system of thought." He sees, from this somewhat positivistic point of view, the "logically uniform system of thought" as the final goal of his attempt.

Einstein obviously understood that the "chaotic diversity of the sense-experience" implies the sense-experience of a three-dimensional space, and that this limitation stood in his way to build a "logically uniform system of thought." The great impediment was the constant velocity of light, and consequently he had to extend the three-dimensional space to a four-dimensional one, which transcends our "chaotic diversity of the sense-experience." Taubes's six dimensions, which he obviously regarded as not belonging to the dimensions of reality, are thus to be augmented by one more dimension—Einstein's fourth.

Recall the idea that the subatomic structure of space-time (the matrix of our present cosmos) is "foamy," originated with John Wheeler (1950). Wheeler proposed it as a means to avoid the absurdity of the actual infinitesimalness of space-units and time-units, a problem that Einstein did not solve. Wheeler described thus the microstructure of space-time in terms our limited faculties of senses and intellect could comprehend.

The classical big bang model contains the flaw that creation of super massive particles would cause the young universe to collapse again after 30,000 years. Alan Guth tried to smooth out this inconsistency and proposed an "inflatory stage" lasting from 10^{-35} until 10^{-32} seconds after creation—a very short extra surge of expansion. Whatever expanded was regarded as one of the bubbles in the tiny drop of froth (just 10^{-33} cm. in diameter) and the first appearance of Wheeler's foam. Only one inflated bubble survived to become our universe. All the competing bubbles collapsed again.

We are permitted to regard this model proposed by Alan Guth as one among many other legitimate speculations about the origin of our cosmos; but soon after there emerged some less legitimate ideas. The confusion may have begun with the cryptic remark by Alexander Vilenkin, shortly after he fled to the United States. "What is Nothing?" he asked. "Classical space or classical time are something, but Nothing is a state without classical space or time—space-time foam." If Vilenkin would have meant that space is perceived through our space-sense and time through our "time-instinct," hence that both are perceived as real, or somethings, I would agree with him. But this fact does not turn space-time foam into "nothing." Space-time foam is still composed of two somethings, space and time, which were created out of Nothing.

Then came Richard Gott who asked the forbidden questions about times prior to Creation and about the Nothing in which our expanding universe is imbedded. He called this hypothetical matrix enveloping our cosmos "superspace," and he endowed this new medium with various hardly credible attributes. It had to create, through eternity, an actual infinite number of bubbles—most of them collapsing again and some of them growing out to become full-grown universes.

This superspace had to expand at a tremendous rate; otherwise it might happen that bubbles too close would merge, a process that has never been observed, at least not in the bubble that is our own universe. We define froth, foam, and the like as bubbles of low density generated in a medium of high density, and Gott borrowed this fact of physics to apply it on his superspace. He gathered that the density of the matrix must be tremendous, something like 10^{93} grams of mass-energy in every thimbleful; and it must be tremendously hot, possibly 10^{32} degrees.

And finally, we must suppose that superspace is unbounded, for if we would surmise that its size would be finite, we would beg the question in what sort of medium superspace would be imbedded in its turn, and so forth. Gott proposed in fact that superspace generates an actual infinite number of bubbles, and this implies that superspace is unbounded.

Gott's procedure recalls the reasoning of theologians who conceded that God has no attributes, and then they proceed to describe Him with all the superb human virtues that cross their minds. Especially obnoxious in his introduction of the actual infinite in his calculations, Gott applied pure mathematics with its unlimited possibilities. His superspace is either actual physics, and cannot be actually infinite in size nor eternal in time, or it is just the outcome of mathematical calculations—not actual—and that is nonsense. Nothing is just nothing. It has no temperature nor identity and does not expand at a certain rate. All these arguments may suffice to reject Gott's model.

If you hesitate to accept my reasoning I may convince you with the following experiment. Let's go back to the first chapter and modify the content a little. Let us suppose that Gott's superspace has never been created, that it has an eternal past, and let us keep a moment in the infinite past in mind. An infinite number of years have to elapse after this moment before the tiny bubble would germinate, which should become our universe. Hence our universe would never have been created. Created out of what? From superspace? It seems that Richard Gott took the first step on the forbidden way of the infinite regress.

We fall into the same trap when we scrutinize the other parameter of superspace, its extent. Physics forbids the actual infinite of the dimensions of a concrete body, superspace, which is supposed to be an actual (physical) reality, and it must be finite as to its dimensions. And the next question is of course: In what kind of medium is superspace supposed to float? In a super-superspace?

Richard Gott solved nothing because of our human impotence to solve the Nothing.[1]

I would have been guilty of a serious omission if I would have failed to relate this subject on space-time to religious philosophy. The Greek philosopher Strato of Lampsacus, who died c. 269 B.C, was what Anthony Flew called a "radical naturalist" who maintained that all the phenomena of the universe can, and must, be explained without any reference to any principle, or principles, in any sense outside or beyond. We may call this the Atheist Argument.[2]

This view has been defended by the followers of Aristotle who was convinced that nature is a brute fact without a beginning and without an end. Natural laws are, so they maintained, dove-tailed in such a way that all the processes are arranged along a series of causes and effects in a closed system.

But philosophy, old and new, and the many youngest factual discoveries refuted the views of Aristotle.[3] Time has been created and this follows from the sound CA of the Kalam. Space-time has been created and this follows from Einstein's theory of general relativity. These two theories derived from different premises, have been corroborated by the discoveries of a number of facts.

Those who would nevertheless stubbornly maintain the view of Strato (called by Flew "the Stratonican Presumption") would be compelled to believe in the spontaneous coming into being of the world in its creation without a cause. And here I have to appeal to their common sense, to everyone's innate conviction, that change and the extreme case coming into being, which would not be the effect of a cause, is nonsense.

We have seen earlier that Bertrand Russell defended the Stratonican Presumption until his last breath. "If everything must have a cause then God must have a cause. If there can be anything without a cause, it may as well be the world as God." This is not a valid refutation of the First Cause, but of a popular sophism, put into words by Flew as follows: "Since everything must have a cause, and since the series of causes could not go back indefinitely into the past, therefore there must be a First Cause, in the beginning." This sophism contains one serious mistake and one omission: The mistake is the premise, "since everything must have a cause does everything include God." The correct premise is that we are all endowed with an innate conviction that every change, which includes even coming into being, must be the effect of a cause.

The obvious omission is that the sophism ignored that effects are consequences of causes. It ignored our innate conviction that the arrow of time flows, points, in the direction from the causer to its effect, and that this implies that a time-interval links the cause and the effect. Our very life is built on this conviction, which presupposes the existence of time.

I remind you that we concluded that time must have come into being, and that we based this statement on the obvious truth that a day in the infinite past cannot exist. Let us call this statement a. I just mentioned our innate conviction that every coming into being must be the effect of a cause. Let us call this statement b. From a and b follows that there must be a cause, that time came into being, let us call this statement c. But we have just seen that causes cannot lead to their effect without the linking time-intervals and that time-intervals presuppose the existence of time. Let us call this statement d. From c and d follows that the course that brought time into being must be the First Cause linked to the first effect by a time-interval. It is a truism that we cannot imagine a world without time and another truism is that theologians usually define God as the First Cause that created the world.

This reasoning clinches the argument and eliminates a common error that has been scorned by Flew: The premise that "everything must have a cause" is erroneous. We may ask theologians who fall into this trap, does God have a cause? And we cannot allow a counterargument that God is not a thing. The tricky word "everything" would indeed include the First Cause, and the false statement would imply that the First Cause has a cause, which is self-contradicting.

Flew added a short account about the causes of motion as a central theme throughout the history of philosophy. It contains a small misunderstanding, and his words are too succinct to be understood at first glance. I have to expatiate. Aristotle and generation after generation of his followers, Thomas Aquinas among them, could not comprehend that moving objects keep moving without being pushed all the time by another moving object. They reasoned that this other moving object is being moved by a third one and so on until one arrives at the First Mover—functionally responsible for keeping the universe going. Christianity codified Aristotle's world by adding the doctrine that this First Mover is God who created the world.

One has to infer from Flew's explanation that Newton's first law of motion would have put an end to the Aristotelian prejudice, but this is not exactly the case. Newton's First Law of Motions says every body continues its state of rest or of uniform motion in a straight line, unless it is compelled to change that state by forces impressed upon it. The meaning of Newton's words are often misinterpreted. I quoted in this chapter Newton's own comments. He rejected the idea that one body may act upon another through a vacuum.

Newton's discovery did not upset his faith in the mysterious faculties of the imperceptible ether to transmit movements. Newton's laws marked only the incipient stage of the great revolution. The real one came later with the abolition of the ether by Maxwell and Einstein. There was not even a substance that transmits movements from one body to another; bodies kept moving without being pushed by other bodies and without the intervention of an interposed medium. The Prime or First Mover of Aristotle became obsolete and redundant.

But in the eyes of Christianity, Aristotle was nearly as holy as the Church. The Church's religious philosophy was based on a modified version of Aristotle's Prime Mover. Theology had to start from scratch after the paroxysm in thinking set in motion by Newton. We owe much more to the Kalam, to general relativity and its consequences, and finally to the corroboration of this theoretical evidence found by modern astrophysicists for having laid the cornerstone for a new and much more convincing religious philosophy.

Modern physicists live unfortunately in a topsy-turvy world. They ignore that our sense perceptions are bound to deceive us. They ignore too that the two parameters, space and matter, belonging to the gestalt of our physical world, are suspect because they are the outcome of data received by our clumsy senses and interpreted by our limited intellect. But we feel that the parameter time is by no means imaginary. Time does not emerge from a sense organ but from a deep-seated instinct. Stephen W. Hawking (1942–) author of *A Brief History of Time: From the Big Bang to Black Holes* (1988) tries hard to translate the reality of time into the rather make-believe world of length and mass (space and matter). He does so by using "the Richard Feynman Principle" that tried to solve the wave/particle duality of light with the aid of quantum-mechanics introducing the "sum over histories" of a particle, accounting for Niels Bohr's allowed orbits of electrons. This manipulation involves higher maths and this subject doesn't belong in the framework of my simple account. Small wonder that in Hawking's mind where linear and mass-units are real, time becomes imaginary—a result my intuition rejects.

Notes

1. On the views of John Wheeler, Alexander Vilenkin, Alan Gush, and Richard Gott, see Marcia Bartusiak, "Before the Big Bang: The Big Foam," *Discover*, September 1987.
2. Anthony Flew, *An Introduction to Western Philosophy* (chapter 7).
3. I refer to chapter 2 and 3.

11

Immanuel Kant and His Spiritual Inheritance

The tenet that the external world is of the nature of consciousness is the central principle of the so-called neo-Kantian Idealistic School. It is an idea going beyond Immanuel Kant's intentions. Kant was not an idealist. He always refused to speculate about reality beyond our sense perceptions. He only maintained that it is erroneous to assume that existence is being perceived through our senses as it is. Kant was a non-materialist, but he believed that we mortals are unable to lift the veil that conceals reality behind our perceptions. We went a step further than Kant and concluded that the world of our senses is a reflection from a mental reality. I refer to the chapter on the mind-body problem in part 1.

Kant is not regarded by everyone as the great thinker of his time. Bertrand Russell mounted his own hobby horse to attack him. "Kant's arguments do not matter here," Russell wrote in his outline of philosophy. "Kant has the reputation of being the greatest of modern philosophers, but to my mind he was a mere misfortune."

My criticism is much less severe, but here follow some of my own objections. Kant believed that our mind is an ideal machine, and that it mirrors, without any distortion, the absolute causality of the external world—though its real character, the *an sich*, can never be known. He was not aware that more than seven centuries before him, great medieval thinkers had shown by the most simple reasoning, that time has been created and that this forgone conclusion, and its consequences, militate against the laws of casualty. Nor could Kant have been endowed with the eyes of a prophet, and we cannot blame him that he could never have believed that the medieval discoveries would be corroborated a few centuries after his life, and that not only time but also space must have been created.

Let me quote Kant's own words:[1]

Now if I inquire into the magnitude of the world as to space and time, it is impossible, as regards all my concepts, to declare it infinite or to declare it finite. For neither assertion can be contained in experience, because experience, either of an infinite space or of an infinite elapsed time, or again, of the boundary of the world by the void space or by an antecedent void time, is impossible."

What did Kant mean by "all my concepts?" The concepts of his eighteenth-century mind? If he meant mental experience, he only underscored his ignorance of the Kalam mental experience—the conclusion that space and time have been created. That the possible infinity or finiteness of space and time would not be "contained in experience," is yet another mistake. Though Kant could not have foreseen that in the twentieth century, experience would prove that time and space are finite.

Let me, in addition, point to Kant's unfortunate wording of his explanation. Time is not "preceded" by an "antecedent void time." Absence of time is not "antecedent void time," nor is the world bounded by "void space," but by the absence of space.

His "second antinomy" is not less confusing and his explanation is confusion with a vengeance. He regarded the question if "bodies in themselves consist of an infinite number of parts or of a finite number of simple parts," unanswerable. Kant made himself clear about his use of the term "simple parts." They are the smallest fragments that have the same essence as the undivided whole, and still smaller fragments have a different essence. Kant is therefore talking about what we call molecules. Experiments, in other words, experience gained from our sense perceptions, have taught us in the nineteenth and twentieth centuries that bodies consist of a finite number of simple parts. The alternative, the statement that bodies in themselves consist of an infinite number of simple parts, has been found to be erroneous.

But this discovery should have been foreseen by Kant as would follow from his *Critique of Pure Reason*: Nature does not consist of dimensionless points and bodies are not composed of an infinite number of points.

Kant's philosophy suffered from further antinomy. He suggested that there exists an unsolvable conflict between two statements that are as far apart as cheese and chalk: freedom of will as opposed to the determination of causal processes. Kant was well aware of the categorical difference between the "ought" of a moral obligation—or the "ought" of an advisable action—and the "must be" derived from logical and causal reasoning. Kant even wrote two separate books, one called *The Practical* and one about pure reasoning. The one was written on how we "ought" to act and the other on how we "must think." And yet he confused common sense with moral sense, which does not contribute to my trust in his philosophy.

We should however be grateful for his positive contributions. We are only requested to exercise some patience in following him step by step along the road of exact reasoning to the bitter end of a wrong conclusion (because his premise was wrong). In order to realize how much we owe to Kant in his methodical mode of procedure, not only his highbrow philosophical speculation but also in the humdrum of scientific research.

As to his precious legacy I mention two points:

1. the sharp distinction of *a priori* laws of thinking from *a posteriori* conclusions; and
2. the sharp distinction of analytical from synthetical judgments.[2]

Kant's lucidly formulated rejection of St. Anselm's OA was a pinnacle in the thinking of the eighteenth century. But Kant committed serious errors in his dealing with the CA and AFD. He started with the wrong definition of God, Whom he regarded as the Supreme Being, according to the custom of his time—a Person endowed with an actual infinite number of divine perfections. We have seen that this definition is a non-starter. God defined as the First Cause—a definition that practically every theologian accepts—cannot be a Person because the First Cause is the Nothing that creates people. Nor may the First Cause be endowed with divine perfections for who may have endowed Him with the perfections He created? And even if He would possess attributes of perfections how could their number be an actual infinite? A potentially infinite number is not possible either since this means becoming infinite, which is never reached, and this would mean that the number of God's perfections would increase in the course of history. How could this be compatible with the timeless Nothing?

Let us now have a look at Kant's concept about the AFD. Though I agree with his point of departure, he defended this argument for the wrong reason. We cannot agree with his idea, which he probably borrowed from Thomas Aquinas, that God moved the inanimate world by His higher will. Kant knew of course that the inanimate world follows strict causal laws, the laws of nature. The rules allowed by the biological world were, in his mind, something else. "Is it not true," so he reasoned, "that organisms and organs are means to an end? Is it not clear that they have a function to carry out, a task?"

I am afraid that we encounter here a serious misunderstanding. Kant may have ignored an essential element where "purposes" are involved. He believed that there must be a conscious "somebody" who has this purpose in view, and that this conscious "somebody" must have planned the organ in its smallest details. But Kant was aware that his "conscious somebody" cannot be the organ itself. My heart performs a function but my heart is not conscious of this task, nor did it consciously develop in order to perform this function. This means that the conscious somebody, which is supposed to have planned my heart, must be outside and independent of my heart. It must even be outside of myself. And this brought Kant to the Great Designer.

I fear that we are entangling ourselves hopelessly in anthropomorphic thinking. The idea that means are consciously used to serve ends is a patently human way of thinking. It is grafted onto our minds as a kind of *a priori* we handle accordingly in every day of our lives. Kant projected this human habit on the development of the animate world when he talked about "the purpose" of an organ.

Means leading to an end and causes bringing about effects are human interpretations of processes. The difference between means leading to ends and causes producing effects is that the ends are reached through a will and that effects are the consequence of a natural law. But where is the line of demarcation? Does the demarcation lie at the stage where the animate is separated from the inanimate?

This question is exactly the key to answer the major problem. The supposed demarcation-line between the animate and inanimate becomes more and more blurred with the progress of the bio-sciences. If we keep this in mind and realize in addition that organs appear through our senses as physical objects, a matter which is the sense perception of mental reality, we are compelled to conclude that the demarcation-line is nothing but fiction. We may assume the position of the demarcation wherever we want, at whatever point we may arbitrarily decide, along the line, running from the most inanimate to the highest animate. We will always notice that the section on the inanimate side is not less mental than the section on the animate side, but the two ends of the line are not reversible. At the "lower" end we are in the realm of the unconscious and at the "higher" end in the realm of the conscious. Consciousness, however, cannot be used as a sharp criterion.

Consequently, events at the conscious, animate end look like the pursuits of a purpose, and events at the unconscious inanimate end as the obedience to a causal law of nature. But this distinction could certainly be nothing but a product of human, fallible imagination—the two aspects of one and the same principle. The essence of this principle may be a secret that may never be unveiled for us mortals. I believe that we are well advised to change our vocabulary somewhat in order to avoid further confusion. We should use terms like "purpose," "means," and "ends" only in the context of the human will. Terms like "cause" and "effect" should refer to the laws of nature.

There was no such distinction in the primitive and the naive mind of the evil thinkers. St. Thomas Aquinas argued as follows in his *Summa Theologica*: "Now whatever lacks intelligence cannot move towards an end, unless it be directed by some being endowed with knowledge and intelligence; as the arrow is shot to its mark by the archer. Therefore some intelligent being exists by whom all natural things are directed to their end, and this being we call God."[3]

The state of mind during the Middle Ages was the belief that inanimate objects have no intelligence, but are used as means to an end by the Supreme Being. So-called laws of nature did not exist yet. Immanuel Kant understood that the inanimate world is guided by the laws of nature, nevertheless he persevered in applying St. Aquinas's AFD on the animate world.

Kant considered the human mind as an unchangeable soul. It was the traditional way prevalent at his time—the mind was a little angel with wings—capable of grasping the truth. Things like mental development and psychogenesis are ideas of the twentieth century. He did not pay much attention to the possibility that all our efforts to interpret some of our sense perceptions may inevitably lead to conclusions we regard as absurdities. Nowadays these absurdities have become the principal problems of philosophy. And this is exactly why it is impossible to apply Kantian-Newtonian philosophy on modern science.

Let us summarize this chapter with a synoptic view of the history of the rise, the glory, and the fall of Newtonian-Kantian philosophy. The Newtonian-Kantian concept of space and time prevailed for so many years—until the second decade of the twentieth century—because of man's indestructible belief that his innate, subjective, unalterable convictions express the objective truth about the external world.

The late nineteenth century French mathematician and scientific philosopher Jules-Henri Poincare stated in 1905 in *Science and Hypothesis*: "Have we any right, for instance, to annunciate Newton's Law? How do we know that this law, which has been true for so many generations, will not be untrue in the next? To this objection the only answer you can give is 'it is very improbable.'"

The next generation came and the very improbable became a reality. The Newtonian-Kantian concept was not good enough and Albert

Einstein replaced it with the principle of relativity. But a great many modern physicists are even today only vaguely aware that the development of modern natural sciences did not just mark the stride of progress, but that with the advent of this new system one had to realize that our obstinate *a priori* on space and time must have always been out of step with the ultimate implications of our sense perceptions. Man could not have been aware of it before mathematical science created the tools that Einstein could use to formulate his theory of relativity, and before experimental physics and astrophysics invented the tools to corroborate its truth. This momentous stage in the history of science unchained an open war between our basic causalistic laws of thought together with our innate perception of space and the ultimate consequences of the interpretations deduced by Albert Einstein of our sense perceptions.

I coined above the term "the causalistic law of thought"—defined as our *a priori* conviction that the law of cause and effect (also called the principle of universal causation) is valid both in the trains of our private thoughts as in the external world, in its processed, changed events. However, conviction that something is true does not prove it is true. The strength of our conviction implies that we are nonetheless unable to conceive of changes without a cause, that we are simply born that way. Our life—language, behavior, and so forth—is conditioned to comply with this innate causalistic thinking. The idea "effect," as the final stage of a process, cannot be conceived without an engendering cause. We feel that spontaneous effects are as abused as effects without causes.

One is hardly aware that the principle of universal causation rules the cosmos; and the factual ruling of the cosmos by the principle of universal causation has been obtruded upon us in the twentieth century. An *a priori* notion was, in the times of Kant and Newton, tantamount to objective self-evident truth. Today, however, we are suddenly overawed by the great miracle, mentioned in chapter 2, that our causalistic conviction is so strictly applicable to experience in everyday life, and though the real cause of a known effect is not always so easily deducible, the external world runs its way in accordance with our causalistic expectations.

The awareness of this dualism, between our expectations on the one hand and the actual processes on the other, with the two running parallel, awoke with the emergence of modern science. The principle of universal causation had to be revised after the physicists succeeded to reach down into the realm approaching the nonentity of the actual infinitesimal; that is, into the realm of the subatomic world.

In the subatomic world the same cause may engender one among several alternative effects, and one cannot predict which effect will be realized. But, nevertheless, even in this world of ultimate smallness there is no arbitrary spontaneity. Causes still engender effects. Even the creation of the cosmos is in need of a cause...the First Cause.

From ancient time until long after Kant there was only one universal principle of which we have an *a priori* notion, a self-evident truth, "the principle of causation." The first Greek thinker who formulated the principle was Leucippus, mentor of Democritus. His teaching that nothing happens without a cause and that every event happens of necessity is vouched for to this day. But our discovery that in the subatomic world a cause may make an arbitrary choice among several effects has become a touchy spot, because our paltry intellect refuses to accept it. Relativity can never satisfy our *a priori*. We will always be faced with deductions and observations that leave us dumbfounded. We are conditioned that way, we are born dumb and blind.

Albert Einstein defined science using the words "science is the attempt to make the chaotic diversity of sense experience correspond to a logically uniform system," and not "a causalistically uniform system." This must be clear when we realize that Einstein included time in a fourth dimension, which is as linear and directionless as the other three Euclidean coordinates, a step that meant that our innate experience of time had to go by the board. Time, though, is experienced by everyone as the vehicle on an irreversible course of the causes leading to effects.

But Einstein had to leave the last domain of our inner convictions intact. It is the domain ruled by the laws of logic. If we would feel obliged to give up logic too, we would step beyond the verge of talking sense. Let me recall earlier how we are conditioned as human beings with innate notions on space and time.

Our space-sense, which physiologists locate as two symmetrical sense-organs near our inner ears, obtrude on us Euclidean three-dimensional space. Its function is senso-motoric, which means that we perceive space as soon as we activate our muscles that move our head, our body, and our individual limbs. Space-sense provides us with a three-dimensional frame of three principal directions: (1) upward/downward (the vertical); (2) sideways (the lateral); (3) forward and backward (the frontal dorsal). We call these three principle directions the three directions of space. All the other directions that we may produce senso-motorically by the movements of our limbs are definable as intermediate (polar) directions between these three principal directions.

These facts become understandable to us when they are brought to our attention during high school. In the later years of our education we learn that René Descartes simplified this system with a bold step. He passed three mutually perpendicular planes through the framework of the three rectangular Euclidean coordinates so that the position of every point in space could be defined by three distances—the lengths of the three perpendiculars from the point on the three planes.

In the mind of Newton and Kant this was the only space there is and from their point of view they were right. Euclidean space describes exactly what everyone intuitively feels about space. We are not taught Euclidean space at school; it is enough that the teacher makes us consciously aware of it.

Kant called space and time, as they are defined by Newton, "*a priori* particulars." He failed however to realize that these two particulars cannot be dealt with on an equal footing, that we receive our space perception through one of our senses, but that we cannot point to a sense organ that provides us with the perception of time.

Let me recall how Newton defined space and time: "Absolute space, in its own nature, without relation to anything external, remains always similar and immovable. Absolute true mathematical time, of itself, and from its own nature, flows equally without relation to anything." We also all feel today that this is a proper description of space and time as long as we confine ourselves to subjective time and senso-motoric (plus visual) space. Subjective time and the causalistic laws of thought belong together. It was unthinkable in the time of Newton and Kant that the external world would not be the obedient servant of these two *a priori* principles.

Then came the nineteenth century with the new constructions of the mind—mathematical tools that enabled Einstein to launch, a century later, the principles of relativity. Jules-Henri Poincaré happened to live and to work in this interim period of turmoil and confusion in the world of mathematics preceding Einstein's time. It explains his confused concepts. He made the fatal mistake of calling Euclidean space "a convention." And one time he remarked: "There is no absolute time. When we say that two periods are equal, the statement has no meaning, and can only acquire a meaning by convention." It is nothing of the kind. Three-dimensional space is, as we have seen, a senso-motoric, innate frame of reference to reach an object. Time, however, is always subjectively felt as the irreversible succession of instants, intervals, durations.

It is our innate feeling that time flows at an unalterable rate; we call a spade a spade; that is, a day a day, or a year a year, as intervals of measurable objective and unalterable duration. When we say that two durations are equal, we feel them as equal and they do not acquire their equality by convention. This unalterable conviction has nothing to do with the illusion we feel that time rushes by when we are engaged in pleasant activities and creeps when we are bored.

In such cases we actually feel that we are being misled. The new contrived geometrical system of Georg Friedrich Riemann (1826–1866) and Nikolay Lobachevsky (1793–1856) were in Poincaré's time still regarded by many as theoretical curiosities, and they did not yet upset the tacit belief in a perfect consonance between our conditioned faculties of perception, our logic and causalistic *a priori*, and the interpretation of the data our senses receive from the external world. But then came the discovery of the absolutely constant velocity of light that upset their complacency. The general public is not aware that this discovery played havoc among the most common parameters. I have to illustrate this with a modification of the simple description of Nigel Calder, which should convince everyone.

A schoolboy has been instructed to calculate the velocity of a ping-pong ball from a certain setting. He is supplied the following information: a ping-pong table is placed in an open train coach; its length at a right angle to the train coach. Two players warm up as the train leaves the station. The players start playing, and at a point along the route the velocity of the ball moving (to and from) between the players is X and the constant speed of the train is Y.

Question 1: What is the compound velocity of the ball from the viewpoint of a stationary onlooker who observes the game looking downwards from the center of a bridge over the railway tracks? Question 2: What is the velocity of the ball from the viewpoint of one of the players?

It is clear that the ball moves more or less in a straight line, at a right angle to the motion of the train in the eyes of the two players, but the onlooker gets the impression of a zigzag line because of the motion of the train. This line is longer than the total distance covered by the ball; and for the two players in the coach. Hence the velocity of the ball is greater for the observer from the bridge than for the two players.

Another setting: a stationary observer is looking upward at two space ships traveling one alongside the other at the same high speed. The captains of the two crafts exchange a signal between them, a light beam reflected by a mirror (primitive means of communication for space-travelers, but for the argument's sake). Again the observer will be looking at a zigzag line covered by the luminous beam. The space travelers, however, see nothing but a straight ray at a right angle to the line of light. In this case we are not permitted to instruct a pupil to calculate the velocity of the light along the zigzag line because of the discovery that the velocity of light is independent of the velocity of the two spaceships.

But now the setting becomes really interesting. The observer measures the zigzag line and finds it to be longer than the to-and-fro one observed by the passengers aboard the two ships. But it equally becomes clear that the moments of the start and the end of the exchange of the signal are the same for the three. Reason: let us take stock of the two traveling UFOs. The velocity of light equals a short path divided by a certain time-interval, and for the onlooker the same velocity equals a long path divided by the same time-interval. Hence, logic cannot be restored without tampering with the length of the centimeter (a way to solve the quandary as proposed by H.A. Lorentz) or with the duration of the second (as proposed by Albert Einstein). In our case either the length of the centimeter has to be shortened or the duration of the second has to be increased for the observer standing on earth.

The given subject matter, as Einstein called the sense experience, is that the constancy of the velocity of light is an unsolvable matter of fact; and it turned out that all the subsequent investigations have justified Einstein's choice, his decision that what had to give in was the duration of the second. Our innate conviction that the units of time-intervals have an absolute duration is only valid from our subjective time. It breaks down when we compare the duration of our seconds with the seconds of our visiting aliens who move at great velocities in relation to our position.

The dependence of durations on velocities is of course mutual—the roles are interchangeable. We may alter the setting by, for example, proposing that a solitary space traveler looks down on my toying with a light-signal reflected by a mirror at a right angle to the direction of the spaceship. The ship's passengers will see my light-beam again as a zigzag line, and exactly the same paradox emerges.

Let me restate: the deduction and corroboration by practical science that duration depends on relative velocities is not just difficult to comprehend, it is mind boggling. It is, in view of our subjective *a priori* of absolute time, simply unacceptable. To us a second is a second, but our *a priori* particular is at loggerheads with our sense perceptions. In simpler words, we could not believe our eyes. This became a moment of deep crisis in the history of science.

Let us regard the consequences when we take the actual mental world into account. Our aging is a cruel process we directly experience; that is, not through our senses but directly in the performance of our mind. It is a process obtruded on us beyond our will, in fact against our will. Could we retard it? Ovid once sang: *O, Mihi, praeteritos, referat si, Jupiter annos*—which means, O, may Jupiter return to me my years of the past. Jupiter never answered Ovid's prayer, but there came a latter-day god named Albert Einstein who could have performed this miracle given the right equipment and tools. He would have put Ovid in a very fast spaceship and hurled him for ten years into the interstellar world. Let us suppose he did. Ovid's twin brother (supposing he had one) stayed on earth looking at the red-shifted tail-light of the spaceship.

Ovid observed exactly the same rate of redshift in all the lights burning on earth. But a strange thing happened ten years later after Einstein signaled that Ovid should return to earth. At that very moment Ovid noticed suddenly that all the lights on earth became blue-shifted. But his twin brother observed nothing of the kind. It took Ovid another ten light-years before the signal from Einstein to return could reach him, and during all the agonizing ten years the tail-lights of Ovid's ship still looked red-shifted.

At this point of my account I have to add a fact from general relativity. If we see a redshift of a disappearing spaceship, we would, if we had the means to zoom into the control panel of the spaceship, see that its clock slows down. When we see a blueshift of an approaching light of a ship, time on that ship appears accelerated and we notice that it has faster clocks.

When the trip was over Ovid received the red carpet and a warm welcome home from the old Professor Einstein. But what a disaster it was for Ovid when he looked at the haggard, old and decrepit friends and his brother. Everybody looked on Ovid's youthful countenance with envy. Ovid consulted his own calendar with the calendar on earth. It showed a day many years later that his.

Did he feel that Einstein had given him back his years? Had Einstein at least answered his prayer? Not in the least. He does not remember to have felt any event that justified this assumption, nothing which could have affected the rate of flow of his subjective time. The big difference only emerged after he compared notes with his old friends. Only this final check proved that the aging of Ovid had been retarded.

But it turned out to be of little comfort to him. It did not bring back his past years, and it did not affect his Newtonian-Kantian subjective *a priori* particular. It did not concern his subjective experience of the unchangeable inner mental clock that continued to mark absolute true and mathematical time that flows by itself and from its own nature equally without relating to anything.

But there was of course one great difference between the vicissitudes of the home-comer and the ones that elected to stay on earth. Ovid experienced several sharp jolts when he was hurled into space, when he had to return, and when his ship braked to land on earth. These jolts were in fact acceleration, and translatable into the effects of strong gravity fields. (Nigel Calder commented that the best remedy to keep young is to travel in a tight loop around a black hole.)

Though it is difficult to see how all these apparent rejuvenation cures could be put into practice, it is all sound theoretical science; and this compels me to look at it from the vantage point of mental reality.

Is aging a physical or a mental process? It is of course a question that should never be asked; but I did. There are no actual physical processes. There are only mental processes. Let me refer to the brain specialist in Part I who had to deduce from his observations of a thinking person, that he could observe from his thoughts only those aspects that he could receive though his instruments and senses; and which he can only interpose as brain-processes. In other words, as some arbitrary physical processes taking place in a gestalt world of space and matter—a shadow world that hides an actual world of mental reality of which the thoughts of the observed person are part and parcel. Let me make it once again clear that I am talking about an external world of mental reality, not about the mental character of our perceiving and interpreting of sense data.

Let us imagine that Ovid had a blind companion unable to perceive redshifts and clocks. Do you believe for one moment that his infirmity would have prevented the retardation of his aging? Of course not; he would have gained as many years, days, and seconds as Ovid. Does this not indicate that the delay in Ovid's aging has little to do with his personal sense perceptions; that it is not his impressions, but an objective reality with a cause in the actual external world? Someone may adduce a counterargument that we have just established that the only effects that Ovid experienced (his friends at home, not) were accelerations. And are accelerations not sense experiences? Take the pressure felt by our tactile-sense in, for example, an elevator. Let me put this person right with the following illustration. Let us suppose that Ovid's fellow traveler was not only blind but fast asleep at the moment of jolts and accelerations. He would have not felt anything of this process that kept Ovid young. But would his state of temporary oblivion have any influence on the rate of aging of the blind passenger? Of course not; but let us go on. What about the retardation of the clocks in the spaceship compared with the speed of time outside? Is a clock not the most inanimate object? We cannot suspect a clock of having sense experiences.

What is therefore our conclusion? Though the duration of a time-interval depends on the velocity of a moving clock compared with the velocity of another clock, there is nothing as objectively real as time, of inner subjective time as a real experience as well as objective time in the external world, unregarded if we measure time on the process in the world of physics or on the mental processes in the minds of our fellow men. The existence of space and matter is only surmised from our interpretations of sense perceptions. We had to infer that they actually have an external mental reality. But time cannot be reduced to anything else but time. The world would cease to exist if time would cease to exist.

We have to tackle the problem if the flow of time is reversible or irreversible, but before we do this I propose to have a look at the concepts of Henri Poincaré about "space."

Though Poincaré called Euclidean geometry "a convention," he conceded at another instance that is was not an arbitrary convention, that it was more a convenient convention, and that any other choice would make geometrical manipulations very laborious. I believe that this correction is nothing but a circumlocution that he had to confess that the term "convention" was a mistake. The real issue is of course not if we would be able to invent as many non-Euclidean geometries as we wish, but which is the geometrical system that described our *a priori* particular of space so that it may be applied in practice. And the answer is, only one, the Euclidean system.

I am, just by chance, looking up at one of the corners of my room, at how the ceiling and the two walls join at three rectangular intersecting planes. Did the builder perform out of convention or are these rectangular constructions intuitively conceived and felt, thousands of years perhaps, before Euclid was even born? I believe that this is a perfect illustration that Euclidean geometry is part of ourselves, and we hold Euclid and his genius in preference because he made us aware of it.

And what about the fourth dimension? Poincaré divided space experience into three different categories—visual, tactile, and motor. I would prefer first of all to change the sequence and to mention first the tactile-motor perception, which I call, after Jean Piaget, senso-motoric space. Neither Poincaré nor Piaget were aware that we perceive senso-motoric space through a sense organ.

Sense perceptions of space through our sense of vision is secondary, space perception through accommodation of our eyes to distance and stereoscopic vision is adapted to the impressions through our primary senso-motoric space-sense during the first year of life. Even an invalid with congenital blindness has a perfect tactile senso-motoric space perception if all his or her other senses are not impaired. But an invalid who cannot activate the senso-motoric organs is in deep trouble.

Poincaré consolidated our visual and our tactile senso-motoric perception of space and called it "Representative Space." I believe that it is not a very fortunate expression. The term "representative" could mean that which can present ideas to the mind, such as imagination, as a representative faculty. I prefer the less confusing gestalt of space represented though our senses.

But Poincaré believed in a second kind of space, geometrical space, which is the object of geometrical science and which allows our mind to wander in an infinite number of multidimensional worlds. Geometry is an abstract science. Only certain domains are applicable on the interpretation of our sense perceptions. Our faculties of abstract thinking, and our faculties of interpreting sense perceptions are so well-attuned one to the other that we immediately feel the difference between the thinkable and the imaginable. We can think about a fourth dimension but we can never imagine a fourth dimension. The above-mentioned congenital blind person is able to mold a perfect cube, but I would challenge the followers of Poincaré who are endowed with senso-motoric and visual faculties of space perception to mold a four-dimensional space model of a supercube of four dimensions. We will never be able to perceive, which is to imagine, a four-dimensional body, though all its properties will theoretically be known.

Poincaré was a man of enormous authority. The influence of his views are even felt today. We find the following comment by Kant in Korner's compendium: "While there is only one finite theory of numbers, there are many geometries among which Euclidean geometry is not necessarily the most useful for the description of the physical world." The problem is much more serious than a simple question of convenience.

Einstein had to introduce a fourth dimension in defiance of our innate Euclidean space, a crucial step which followed on the heels of his decision to modify our innate concepts of time.

The deceptive harmony between our laws of thought and spatial sense perception on the one hand and the laws reigning in the physical world on the other was forever broken. The only realm remaining untouched is our common sense of logic.

And let us now turn to the last subject. Is time reversible? It is clear that this problem cannot be solved anymore with logic. I personally would define logic as the study of those laws of thought—and their application—which are our innate rules of exact reasoning and which are independent of the laws of our causalistic *a priori*. (The law, for example, that the effects follow from their causes.)

Reversible time would imply that the effects would precede the cause and this would contradict our innate notion of subjective time— the *a priori* particular of intuitive experience of an irreversible flow of successive states, and directed into the future. We feel subjective time as the vehicle of process, through which causes lead to effects and where the effects cannot be produced without the preceding, special cause.

If we want to talk reasonably about the possibility if time is reversible, we should thus disregard subjective time, which is irreversible on the account of its definitions. When I encounter in Peter Angeles's *Dictionary of Philosophy* the definition "Time is the irreversible succession of instants," the negation of reversibility that it implies would make this definition unworkable for the solution of our question: may time ever run back? We are not permitted to answer "time is irreversible" because irreversibility is implied in its definition. It would be a stark example of begging the question. The definition expresses, however, subjective time, and we are interested in the character of the flow of time in the external actual world. This problem has been the life work of Roger Penrose.[4]

This work suffers from a slight defect. It brackets subjective time with objective or external time as observed in the physical sciences. This distinction did not yet appear in Newtonian-Kantian philosophy. There was only one time, absolute time. But the realization that subjective time is a mental quality, while objective and external time is experienced as the flow of successions in the physical world, which has to be translated into a flow of succession in the world of mental reality, obliges us to keep the two rigorously apart.

It is, for example, strictly forbidden to reason as follows: a brain is just a computer and the memorizing mind works just like the memory in the computer. The arrow of time of a computer would thus point to the future and not to the past, just as we with our subjective time cannot remember the future and foretell the past. This was the reasoning of Stephen Hawking in one of his lectures, a jumbled and confused mixing-up of the mental with the physical. Even the most ideal computer does not "remember" a thing. It has no more a conscious a memory than a lexicon, or an old-fashioned adding machine, or a ruler, or an abacus. All these man-made electronic devices look like memorizing minds; but in reality they are useful simulators of the physical aspect of a mental function. Active mental memory looks, through our sense, like an active computer, a brain. A computer is not the physical aspect of a mind and it does not remotely resemble the human mind.

The same kind of misunderstanding, though, in a much less offensive degree turns up in Tony Rothman's account of Roger Penrose's *Seven Arrows of Time*. Einstein dismissed the old view that man's feeling that the subjective constancy of the duration of time-units reflects objective truth in the external physical world and this undermines our confidence in our feelings that the subjective arrow of time should reflect objective truth. It is therefore inadmissible to adduce our subjective feelings that time flows only in one direction as an argument to prove that time should also flow in that one and same direction in the physical world. And this is exactly what Penrose occasionally tried to do. The idea alone that time would flow back evoked, of course, the feeling of strangeness of an extreme oddity; but to appeal to this feeling as an augment to dismiss time-reversal in the objective external world is reintroducing the objectionable, subjective time through the backdoor—just after we have thrown it out at the front door.

Another counterargument is that oddity is so ill-defined. There are oddities of many sorts and degrees—from the low-grade oddity, which is recognized as common sense in disguise, to the purely anti-logic oddity, which has to be dismissed out of hand.

Let me start with a low-grade oddity. I found it in the records of the Meteorological Department of the British Mandate in Palestine. It was the following account of rainfall: precipitation on Monday: drops; precipitation on Friday: drops. Summed up for the whole week: three millimeters.

This oddity is the consequence of bad rules or terse recording meant to describe that the quantity of rain that fell on Monday and Friday was too low to be measured separately, but that the totalizer showed at the end of the week a total quantity of three millimeters.

Poincaré mentioned an oddity of the same lower-grade family. A weight of ten grams produces the same feeling in our hands as a weight of eleven grams and a weight of eleven grams feels the same as a weight of twelve; but we feel the difference between ten grams and twelve grams. (Do not try to express this in the terse mathematical equation that $a = b = c$, but $c > a$.)

This kind of oddity is not an absurdity at all. But there are oddities that are really at odds with our common sense, and are nevertheless true. The twin paradox examined above comes very close to such an absurdity because it mitigates against our *a priori* particular of subjective time, which we feel as flowing equally without relation to anything.

But the really stark oddity is, of course, the *Creatio ex Nihilo*. This is (a) an oddity that militates against our common sense; (b) a truth supported by overwhelming evidence, theoretical as well as practical; and (c) it has been purified from possible fallacies, especially of the fallacy of begging the question.[5] One should, for example, never reason as follows: "Everything has a cause. The universe is a thing, therefore the universe is a thing that has a cause." "Everything has a cause" is a somewhat doubtful and unclear statement. If everything means every single item, we are allowed to refer it to the question if every effect has a cause? We have to ask ourselves if everything or every single item is an effect.

Our innate, causalistic *a priori* tends to confirm this and even the so-called uncertainty principle that momentum and position of subatomic particles cannot be precisely determined, nor the unpredictable behavior of a quantum managed to dent our conviction that every single item and its state is a caused effect. But is the universe "a thing?"

Furthermore one should examine if the following reasoning mentioned by Angeles under "Fallacy of begging the question" is valid: "The universe has a beginning. Everything that has a beginning has a beginner, therefore the universe has a beginner called God." Angeles remarks that the question has to be answered if the universe has indeed a beginning and, secondly, if we are allowed to imply that "everything that has a beginning must have a beginner." As to the first question, the irrefragable fact that time has been created implies that the universe has a beginning, because the existence of the universe is unthinkable without time. Furthermore I refer to chapter 1 and 2, for theoretical and practical vindication that the universe has a beginning. The second question is not less interesting. If the term "beginner" should mean a person, we would all of us reject that everything has a beginner. The modifications that everything must have a cause would refer the probable to the one we have just examined: has every single item a cause? And, especially, has the state of every item a cause?

We have, as expressed, strong arguments that this is indeed the case. Our experience from the objective external world has never come to blows with our subjective *a priori* causalistic conviction. But in the case of the beginning of the universe this is not the problem at all. Does the fact that the universe has a beginning imply that the universe is an effect? If it is indeed an effect it must have a cause and therefore an impersonal beginner, which is the same. But if the universe is not an effect, though it has a beginning, it must have had a beginning without a cause, which is the definition of spontaneous beginning.

We did not encounter as yet anything in the external world that points to a spontaneous beginning, an idea which moreover militates against our innate laws of thought. But this is not the whole story. The so-called spontaneous beginning of the universe implies the spontaneous beginning of time. Creation is therefore not the bringing into existence, for bringing into existence would mean that which has been brought into existence came into existence; which is not thinkable without a time-interval. The spontaneous beginning of time would, on the contrary, be more something like the timeless coexistence of not-Being (the *Nihil*) and the Being which is time and the universe. It sounds as if we are transferring the problem from the causal level to the timeless logic level, which leads to the greatest absurdity that Nothing would equal Something. This illustrates that terms like "automatic" and "spontaneous" hide everything and explain nothing.

But the way out is simple. Our efforts to transcend the threshold from the temporal into the timeless have no meaning. The only meaningful conclusion is that there is more behind the *Nihil* in the *Creatio ex Nihilo* than we will ever be able to know. The threshold separates the natural from the supernatural. The premise that everything must have a cause is erroneous. The correct premise is that every thing coming into being must have a cause.

The ultimate aim of my digression is to confront the oddity of the *Creatio ex Nihilo*, with the oddity of time turning tail. The extreme contrast between the two is that we have an excess of evidence in favor of *Creatio ex Nihilo*, and no real evidence whatsoever that would support the idea that the arrow of time could ever point the other way. We may even adduce a strong argument against the idea: what would be the fate of the universe after the very moment when time is supposed to do a U-turn? We may consider either a static or a dynamic state of balance. There is no third possibility. I brought already the consequence of the first possibility under your attention. Time would not even get the chance to turn around if it would screech to a standstill. The existence of the universe depends on the flow of time. But scientists may reproach me with talk, like the grocer who gives clients a fair deal when the pointer of his scale halts at zero.

Scientists regarded a state of balance as a state where the oscillations to the left equal the oscillations to the right. Let us examine if this would make sense in our case at the critical moment—it does not need to be the moment of the big bang, nor any supposed moment of maximum expansion of the universe. Time would run in both directions at once. Some watches may still be scattered on the floor in smithereens, while at the same moment the fragments of other watches may leap from the floor to reassemble themselves into perfect timepieces. On the human level there would be two kinds of creatures, people and retro-people, whose past is our future and vice versa. How would it ever be possible to talk sense to them? When we would have the misfortune to drop our watches, they would be happy that the fragments jump from the floor and reassemble themselves into a beautiful and perfect instrument before we receive it into our hand. But in our eyes they would live in exactly the same crazy world where towels get wet on the laundry line in perfectly dry weather. It is an incoherent world that must fall to pieces in the absence of supporting evidence.

It shows that there is no reason whatsoever to tamper with the arrow of time. It must be a crazy mind that would believe in a world where we carry our wishes in reverse, even before they would be formed in our brain. A world where potentialities would not become realities, but actualities would turn into potentialities (to paraphrase Roger Penrose). And let us challenge Stephen Hawking to compare this baseless oddity with the great oddity, which we are compelled to accept, the Creation out of Nothing.

We may exchange ideas and think hard and long about the source of the hypothesis of time reversal. It may have been the desire of some scientists to assert themselves to look interesting in the eyes of their audience. But Tony Rothman mentioned something else: the Newtonian-Kantian and the Einsteinian models are mechanic models and deterministic. This term expresses that we may predict with cocksure precision what will happen to a system of which all the conditions are known. The exactitude of that prediction will be so great that it would look like the exactitude of a notation of the past written down from infallible memory. Past and future would look in this respect as two equivalents—like the two ends of a finite line. The front page of Tony Rothman's paper is adorned with an hourglass—a proper symbol. When we regard the narrow neck as the present, the upper side as the future and the lower side as the past. And what happens when we turn the instrument upside down? Nothing has been changed, only that the past has become the future and the future the past. Such are the equations of Newton and Einstein. Newton's planets may orbit the sun clockwise or counterclockwise and nevertheless comply with his equations.

In Einstein's special relativity, Euclidean space has been broadened with a fourth dimension, a linear one of course, based on the principle that the multiplication of a velocity of light, in our case by a time-interval, equals a length. A length is not a vector; it has no "direction." The history of an event becomes a world-line symbolizing its movement through four-dimensional space-time. Its past, present, and future are well-defined. Albert Einstein maintained in one of his private letters that, "for us convinced physicists, the distinction between past, present, and future is an illusion, although a persistent one." It is not even an accurate description of his own conviction. The present is a mere point that divides the world line in two interchangeable sensations, the future and the past.

Einstein did certainly not regard time as an illusion, though Newton's equations are time-reversible. According to Einstein's definition, time is not reversible at all. It does not even allow for a change in its rate of flow. Hence, the question "if time could flow backwards..." did not yet emerge. This question only turned up after Einstein showed that the rate of the flow of time was not the unshakable constant that it was hitherto supposed to be. As told, time slows down in the eyes of the fast traveler who looks out of the window, and if we may suppose that he could travel at the speed of light he would observe that the time in the external world would come to a standstill. Our first question would of course be "how did he ever manage to ride on a light beam?" But his prodigious celerity would be the ultimate that relativity allows. That is why Einstein never considered seriously that time could flow backwards.

In short, time has no linear dimension unless it is multiplied by a velocity. But the sign of equality that links the two sides of the purely mechanical equation determines the future with the same exactitude as the past.

But all this science does not describe the functioning of a normal brain. Everyone is aware of his unshakable conviction of both the deterministic and the irreversibility of the flow of time, whether he is a scientist or not. The faculties of our mind and our senses are the essential tools for our survival, but they are not conditioned to discover the truth of the external world. That is why we are unable to grasp relativity, the uncertainly principle, Creation out of Nothing, which seem to violate our inner convictions. I doubt if there is even a glimmer of hope that we may ever be able to solve these antinomies. I fear that we will never be able to perceive the truth that casts the shadow-world of space and matter. The many scientists who ignore that the *an sich* cannot be just the same as what we observe, are like an absent-minded professor who accidentally puts the colored picture under his microscope instead of the original preparation, and then tries to make sense of the incoherent mass of colored dots.

Timothy Ferris asked in his book *The Red Limit* about the matter of wave theory verses particle theory in the description of the character of light. "One would see the subatomic world as waves or particles, whichever one looked for. What if the question of which was 'real' could never be finally decided?" And we have to complete the question of Ferris's, "what if neither would be real?" Wouldn't that not be the most reasonable answer? Is it not obvious that our eyesight cannot perceive reality? And Werner Heisenberg came with the correct answer: "Perhaps we can never observe the 'real' world on such a small scale. If we cannot, then speculation about what the world was really like was not a matter of science."

If it could not be a matter of science, what else could it be but a matter of religion? We however were not yet discouraged so easily that we should confide the problem to the theologians. We consulted the science that deals with sense perceptions and its interpretations, which is the sensory aspect of epistemology (and its metaphysical implications). Let me replay the same old disc: the real world is obviously mental although it is about all we are permitted to say about its essence. This real mental world is related to the world of our senses, as an object is related to its shadow. We have to infer from a thousand arguments and all the evidence that this shadow-world is Creation out of Nothing. Our way and Heisenberg's way meet at this intersection. We are at the end, and any further speculation about what the world really is, is not a matter of science but of religion.

Finally, what about the newest alternative proposed timidly by Stephen

W. Hawking?[6] "The boundary condition of the universe is that it has no boundary." This quixotic remark means that Hawking nipped off the arrowhead from the time-parameter. What remains is just a line indistinguishable from the other three Euclidean coordinates of space. Time has become in this mathematical procedure an imaginary quantity—the square root of a negative one. In other words, Hawking stopped the natural flow of time and the world becomes a four-dimensional globe without any point of special preference. There is no past and no future, no creation and no doom. We could surmise that such a proposal was in the cards after Einstein published his equation of special relativity.[7]

Hawking was well aware that the beheading of time meant the premeditated murder of his own life-work on singularities (dealing with the practical consequences of the actual infinite, such as the big bang and the black holes) for his tentative proposal would smooth them all out.

Let us have a cursory look at my principal objections. His idea runs against our innate convictions, against the reality of the psychological arrow of time. While we feel that the arrow of time in the world perceived through our senses, which is a shadow-world, is less reliable (because the gestalt-world is just a construction of our fallible sense perceptions.) I pointed out in chapter 2 that our expectations how causes lead to effects are always corroborated in what we observe of the external world, which I termed here the "shadow-world," or "gestalt-world" conjured up by our senses. I called this an unexplainable miracle. Why should the external world comply with what we expect? It is a miracle that one never should minimize. Hawking did not mention the principle of universal causation and splits the arrow into two, into the thermodynamic arrow—which is the arrow of increasing disorder in the universe, that is, entropy—and the arrow of the expanding universe. But I feel that these two arrows are special cases of the arrow pointing to causes and effects.

I could have adduced the argument of the double meaning of the sentence, the arrowless time in the external world is imaginary, that the imaginary does not only mean the square root of a negative quantity, but also only existing in the imagination. A model which exists in our imagination and not in reality is suspect. This sounds like quibbling with the double meaning of imaginary, but there is more behind it. Time, the vehicle that leads from causes to effects, would stop flowing and thus cease to exist.

There are other remarks in Hawking's work that are unforgiving. Imaginary time may sound like science fiction, but it is in fact a well-defined mathematical concept. The issue is not whether imaginary time is a well-defined mathematical concept, but if it is a physical concept, which can be tested by observation. And Hawking admitted that his idea cannot be deduced from some other principle, and that there are several reasons why corroboration of his idea would be hardly possible.

And I may answer his remark that imaginary time is a well-defined mathematical concept, with the following objection: Isn't the infinite a well-defined mathematical concept too? But what about the actual infinite? And if the actual infinite does not exist why should actual imaginary time?

Though Hawking admits that his model of a four-dimensional sphere, finite but without boundary, "is just a proposal," but that it "cannot be deduced from some other principle"— it is not independent of another principle. He borrowed it from U.S. physicist Richard Feynman's (1918–1988) clever device to calculate the allowable orbits of electrons around the cores of atoms. It would carry us too far to follow the train of thoughts of these two geniuses, but I feel that their proposals are mathematical means of description and that they fall short of giving a physical explanation. But my principal objection is that such modes do not take the mind-body problem into account.

And this reminds me of the paper "The Mind in Motion," by Geoffrey Montgomery, published in *Discover* in March 1989. Observation of the physical processes in a living brain is not science fiction anymore; it is actually being carried out and the technique is called positron emission tomography. Blood is injected with a radioactive tag that lights up on the spots where the brain is in a state of intense activity. The picture is projected on a screen. One of the directing scientists was Marcus Raichle, at St. Louis. On the blackboard in the conference room adjoining his office, Raichle drew a diagram with two perpendicular axes. He labeled the vertical axis "Space." It described the group's efforts to map the functional structures within the space of the brain. The horizontal axis is labeled "Time" and it described experiments that may help define how these structures are successively activated. This axis leads to an entity that Raichle had labeled "Mind."

Here follows an exchange of words between two of Raichle's colleagues, Peter Fox and Steve Peterson. "Because the brain is a physical structure," says Fox, "it exists in space. But now the mind—the mind operates in time alone." Peterson hesitates, but Fox continues. "Why not? You can only deal with the mind as an entity in time. What other dimension does it operate in?" Peterson calls Fox a dualist, but Peterson would not have used this abusive language if Raichle would have drawn the two lines one alongside the other—the Space line a bit more hazily drawn than the Time line. Whenever new thoughts turn up in the mind, the observer sees other points flashing bright spots on the screen. What is being observed is the only side-effect of thinking, which the senses of the observer are capable of picking up. This information is scanty and its interpretation cannot reflect objective reality, Kant's *an sich* of the mental activity known as "thinking."

Notes

1. Translated from Kant's work *Prolegomena zu einer jeden kunftigen Metaphysik* (1783) by Henry D. Aiken—*The Age of Ideology*. Aiken explained why Kant believed that his first antinomy, the problem of whether the world of space and time is infinite or finite, is unanswerable.
2. For further reading in Kant's epistemological deductions, I recommend S. Korner's *Kant* as a shortcut.
3. The translation of *Summa Theologica* was quoted from *The Age of Belief*, Anne Freemantle.
4. The results are briefly summarized in the article "The Seven Arrows of Time," by Tony Rothman, *Discover*, vol. 8, February 1987.
5. "Begging the question" is defined by Peter A. Angeles as arriving at a conclusion from statements that themselves are questionable and have to be proved, but are assumed to be true.
6. *A Brief History of Time*.
7. See Einstein's *Out of My Later Years*.

12

Johann Gottlieb Fichte, Arthur Schopenhauer, and Max Wentscher

Three Germans, arbitrarily picked out from a multitude. Indeed the only justification to choose these three that may cross my mind is that they are fellow countrymen of the same era. I have placed them in historic order so that I may compare them without falling in gross anachronisms.

I brought you at the end of the foregoing chapter into contact with a patently human faculty that we did not encounter yet, the will. The purposeful urge to accomplish an end by the use of chosen means. Willing is unthinkable without consciousness. A hunting predator, an animal, makes use of its cunning and its other conditioned faculties to catch its prey, but it is not aware that it uses means to an end. Even the terms "means" and "ends" are in this case misnomers in the absence of a conscious will.

This is a reasoning that would have been lost on Arthur Schopenhauer (1788–1860) who objectified "will," making the same kind of mistake as Nachman Krochmal, who objectified the laws of nature and turned them into the independent beings. "Will" became a demon hunting everything in the world, animate and inanimate, even more dominant than in Kant's fantasy—one step further and the organ fulfilling its purpose is moved by a will to use means to an end. But Schopenhauer's war cry—"back to Kant"—had a much wider scope than his agreement with Kant's view on purposes. It was an open rebellion against idealism.

Schopenhauer's obsessive and frenzied "*idée fixe*" regarding will is expressed in the following sentences from his work *The World as Will and Idea.* "A man can also say and must say 'the world is my will',", and "For the world is in one aspect entirely idea, so in another it is entirely will."

He got his cognizance of will from introspection. Will is evil, so he concluded, and this meant of course his own will. Will is evil is tantamount to "Arthur Schopenhauer is an evil man." He might have this printed on his visiting card but we have other ideas about will. Will is neutral from the ethical point of view. The opposite of ill will is goodwill. It depends on the character of the person, if his volition is ill will or good will.

If will is evil it has to be exorcised to be destroyed. But one cannot kill will without killing the self. The only remedy against evil is thus suicide. But Schopenhauer had just enough malevolent vitality to show another way to kill your evil will, by contemplating the world of art and to listen to music. But looking at paintings and listening to music does not kill our will, we are only carried away for a brief moment as we are listening.

Schopenhauer regarded Johann Gottlieb Fichte (1762–1801) as his archenemy. He did not just oppose him. He hated him—and Hegel too. Fichte did not preach solipsism, but he behaved like a solipsist. He did not want to go one step further than to acknowledge that his own sense perceptions are real. He maintained that it does not even make sense to ask if these perceptions are the reflections from something else, from an objective reality. He even dodged a reply thus, "as far as I am concerned, I allow you to believe in a 'real,' though unknowable, world; that is your problem, not mine."

It is, so he believed, a "practical question" not a "philosophical problem," and he preferred to keep to his own opinion on an eventual background of sense perceptions. He was, however, ready to admit that his sense perceptions were obtrusions beyond his will, that "we" all experience these unwilled sense perceptions, and that "we" all interpret these perceptions in such a way that "we" are able to communicate one with the other. "We" means all our fellow people, everyone endowed with his or her own ego. In short he agreed that the world is populated with a multitude of egos. Then all of a sudden he shrunk from the most obvious conclusion: that all these perceptions, experienced by all these egos, must therefore point to a common source, an external world. He remained there, standing on the brink, for a lifetime. His stubbornness was really amazing.

It is clear that our ability to communicate cannot be less real than the reality Fichte was ready to ascribe to his world population of egos. And Fichte should have concluded that his fellow folk must exist just because he is able to communicate with them.

Let us now assume that our ability to communicate is impaired by a wall separating us from others. This wall, which we interpret though our sense perceptions is made of bricks and concrete, and must have some objective reality—insofar as it constitutes a barrier that impedes our ability to communicate. But it is a matter of course that what we all observe as a stone wall is not just the material (as distinct from) object made of bricks and cement in the objective reality, though our senses constrain us to believe that this is really the case. We may on the other hand summon all the egos of Fichte, all the millions that compose what we call mankind, to bear witness to the existence of the wall, and the confirmation by every separate additional witness will add up to the overwhelming cumulative evidence that the wall is not a mere hallucination either.

We may apply the same reasoning to show that space is neither a mere chimera. The farther we are removed from our neighbors the less are we able to communicate with them, the more we have to raise our voices to make ourselves understood. What we perceive and interpret as space must be a reflection of a reality, though the space-sense through which we received this impression certainly distorted this reality. We are convinced that this reality must be of a mental nature; but Immanuel Kant would have maintained that we do not have the faintest idea what the reality is, that which hides itself behind our space perception.

What may be said against Fichte's view is that matter, the wall, and space—the distance—are gestalt expressions that cannot be discounted as products of our imagination. They point to something real in a real external world. To quote Henry D. Aiken: "A plurality of egos, like ourselves, that make common posits must involve the existence of an external world." But Fichte put this in doubt. Berkeley, on the other hand, regarded the existence of the external world as being dependent on the perceiver. Fichte did not even go beyond perception, and if this perception pointed to an external existence was for him an open question, and even a senseless one.

I have one more argument against Fichte up my sleeve. While thoughts and feelings convince us that we are alive, Fichte would not deny that—we are kept alive by food and air. It looks as if food and air, two elements of the worlds of space and matter, are essential for the sustainment of our thoughts, the functions of our mind. So if our mind exists—as agreed between Fichte and us—why should the essentials for the upkeep of our mind not exist in one form or another? Let us poke a little deeper into this truth. That thinking exists is beyond doubt. But that the incongruous element in our argument is how matter can guarantee the survival of mind. How is it possible that the physical world is responsible for the survival of the mental world? Are we not confronted once again with the mind-body problem we discussed in chapter 10?

In short we may counter Fichte with the obvious conclusion that food and air are not mere hallucinations, though this does not solve the real problem of how it is possible that the world of matter interferes with the world of the mind. We have seen that there is only one solution—what we observe as matter is in reality mind. Mind keeps mind alive. The influx of mind (what our senses interpret as food and air) guarantees the survival of our thinking mind.

Did Fichte believe that his nonexistent external world, a world that we only imagine to exist, has any sense? Obviously he did. It is a means to the purpose that we acquit ourselves of fulfilling our duty. And after we have learned that Fichte believed in duty, we might have expected that he would follow Berkeley's footsteps and agreed with him that it is "God's commandment that we do our duty." But no duty is a self-sufficient obligation, and has no higher Master. It was a view that filled Fichte's pupil F. W. Schelling with utter revulsion, as if he was looking in the profundity of hell.

Only a few have followed Fichte along his path of personal idealism; which comes so close to solipsism, the madness of the feeling that everything is a creation of fancy, like the egocentric predicament invented by Ralph Barton Perry, which was paraphrased in the following rhyme:

> If a tree dies in the wood,
> and it falls where it stood,
> and if no one is around
> Will there be a sound

The third figure I picked out at random from the mass of German philosophers who populated the universities one after the other between the eighteenth and twentieth century, was a pedantic scholar named Professor Max Wentscher. He wrote many textbooks before World War I, but he compiled his ideas in a very clear booklet *Einfuhrung in die Philosophie*. He was acquainted with the current views of his time. His explanations were lucid, but incomplete, and that was probably the cause of his serious mistakes.

Let us examine his solution of the mind-body problem. He was of course familiar with the ideas of idealism and parallelism, though he did not associate the one with the other. He did not, for example, explain that the first step in the direction of an idealistic system involves the acceptance of at least a certain parallelism between physical and psychical processes. But he brought the whole principle of parallelism in disrepute by imputing to it the nonexistent rule that every psychical process should be associated with a process in the physical world. This brought him to an erroneous assumption that the principle of parallelism requires that the following three conditions should be fulfilled: (1) the absolute causality of the mental world; (2) the absolute causality of the physical world; (3) the permanent interaction between the two.

Let us return to Plato's den and scrutinize Wentscher's allegations. We have our eyes glued on the wall and the play of shadows from invisible, real objects. Of an absolute causality, which rules the real objects, we are ignorant because they are invisible. It is on the other hand absolutely impossible that the spectators would interpret the shadows on the wall as being ruled by the strictest laws of cause and effect.

Some events of the circus-show would not look reasonable at all. Some of the figures would be pierced by knives and do not react. Large objects are being swallowed down, but the person does not seem to suffer from his greed. But the real objects who are casting all these shadows obey, in their movements, the rules of absolute causality.

The real objects symbolize the mental world and the shadows of the world of our senses. Our senses are few in number and selective in their rather defective performance in picking up information from the external world. If I would be asked to choose which world would follow strict causal laws, I would point to the mental world, the real objects in Plato's cave. That the world of our senses does not follow these rules so absolutely has been discovered in the twentieth century. Particles in the subatomic world behave whimsically.

Wentscher postulated a third condition. The permanent interaction of the physical world and the psychical world to maintain the parallelism. The example of Plato's cave (or the example of the mirror and the clock, Heymans's modification of Geulinck's comparison with two clocks) is an adequate illustration that the mechanism responsible for the parallelism is not a problem at all. We mortals look through the spectacles of our senses at a mental reality and see a physical world.

Our senses are so highly selective that they fail to pick up vital events in the mental world, such as the human will of the people around us, that we have to deduce indirectly from their physical behavior and their speech (equally physical). Senses are only sensitive to those elements in the world of mental reality that give us the vital information necessary to keep mankind alive.

Wentscher adduced a most unconvincing argument to persuade us into believing that the physical world is a world of objective reality. Is it not strange, so he asked, that our senses only observe and perceive the transmission of the physical processes from particle to particle, while the so-called psychical stimulations of these elements are forever unfelt? Does the psychical world not reveal its objective existence, for example, in the fact that we may raise and lower the pitch of a tone, by moving towards or away from the source of the sound?

I do not deny that Wentscher adduced facts of the physical world, but they prove something else...that the external world is a real world existing independently of us, but that it is impossible that the gestalt world we build up from the scanty and distorted information, which is provided through our senses, would be an exact replica of reality.

After this refutation of Wentscher's objections against the form of parallelism—that mental and bodily processes are, in the main, concomitant, as idealism requires—we look for his alternative explanation. And indeed he tried to delude us with an alternative that was possibly the greatest mistake of his philosophical life: the postulate of reciprocal action between the body and the soul.

It was back to Democritus and Epicurus. Physical energy provided by the external world is, hocus-pocus, converted into mental energy and vice versa. We obtained a very valid argument to counter Fichte and we use it here again. We eat, drink, breathe and keep our mental activities going. To interpret this process as the conversion of physical energy into mental energy is not just paradoxical, it is an absurdity that has been expunged from philosophy centuries ago.

Wentscher paid much attention to moral philosophy. He was obviously acquainted with Heyman's principle of objectivity and he defined Heyman's view on moral behavior properly. He was repeatedly unable, though, to adduce valid arguments in support of his skepticism. He called the universal validity of Heyman's principle (he does not mention his name) in question. How so, he said, may we expect a person to behave in a society ruled by club-law, and the law of the fist? Would it not be advisable for him to behave just as his fellow men?

The answer is of course crystal clear if we define properly the term "interests" in the statement that we must regard our own interests on a level with the interests of our fellow men. Should we satisfy the sinful glutton of that guzzler over there? Certainly not. To feed his greed is not a legitimate interest. Interests should first be scrutinized. The interests of our fellow men and of our own interests, if their satisfaction would not harm others or even the health of the aforementioned guzzler. Legitimate interests are bounded by an upper limit. Looking after somebody's legitimate interests means having his welfare, his well-being, at heart.

It is of course perfectly true that Heymans's principle of objectivity cannot be applied under very adverse circumstances, like, for example, when everybody is so much in need that the one gets his bread when the other is dead. Freedom of action has been curtailed by tyranny, in this case by the severe tyranny of famine. The same kind of oppression is felt in Wentscher's example "equal rights were also granted to everyone in the time of fist-law." Indeed, when everyone infringed upon the legitimate interests of others, nobody was free to act. Club-law or fist-law is not law by tyranny. The first condition that has to be fulfilled in these cases in order that Heymans's principle of objectivity may be applied is: restoring freedom of action, in the first case by feeding the entire population, in the second case by imposing laws that enable the members of the society to act according to Heymans's principle of objectivity.

But what did Wentscher maintain with regard to the above? That this freedom of action should be identified with acting morally? Wentscher, again, erred badly. Freedom of action enables one to show his color if he is a good or a bad man. To be free to act as we want does not guarantee at all that our acts will be morally justified. This depends on our character. A good society is, of course, a free society; a democratic society in which the good characters are told to take full advantage of their free acts and in which the freedom of the bad characters is curtailed.

Wentscher's chapter, "The Vindication of the Principle of Freedom," is a dangerous soporific sermon. As a well-known professor in moral philosophy in Germany, he had been an accessory to the permissive politics of the Weimar Republic, which failed to summon Adolf Hitler and to sentence him. It placated him in the name of Wentscher's fake "Principle of Freedom," even though those who bore direct responsibility may never have mentioned Wentscher by name.

The three figures of this chapter—Schopenhauer, Fichte, and Wentscher—are a fair cross section through the various philosophical schools from the seventeenth to the end of the nineteenth century in Germany. All three of them a failure, each in his own way. Schopenhauer was a failure because of his miserable character. Fichte did not have the courage to draw obvious conclusions. Wentscher was a spineless and shortsighted idealist. What has all this to do with religious philosophy, our subject? More than just a little.

The three philosophers in this chapter showed a downward trend in religious commitment to any interest in religious affairs. This downward trend became apparent after Hegel, one of the last German thinkers who used God as the starting point of his system. God figures in Wentscher's writings but he kept within the bounds of a general account at the end of his work. I looked in vain for his personal opinion.

13

Charles Darwin and the Ensuing "-ism"

Charles Darwin (1809–1882) was the counterpart of Karl Marx. They both upset the complacency of the Victorian Age by their outrageous and offensive theories. But Darwin was less a fanatic than Marx. His "evolutionism" did not hit society with the same revolutionary impact as Marxism because it did not preach revolution. Marx acted like a prophet of Divine Truth but Charles Darwin was, in comparison, modest. His works *On the Origin of Species by Natural Selection* (1859) and *The Descent of Man* and on *Selection in Relation to Sex* did not claim to solve problems but to submit problems to the intelligent readers.

It was well-known in his time that the development of the animate world followed some ascending line. It was even, though very roughly, revealed in the first chapter of Genesis. To state that a higher animate form came into being after a lower one would not have angered anyone, not even in Darwin's time. But Darwin claimed that the higher develops out of the lower through the adaptation to changing circumstances. This would not only mean that dexterity comes by exercises, but that the son would bear an innate dexterity, which is the same dexterity that the father acquired by his exercises.

This is of course an unreasonable inference. We may know that new faculties acquired by experience are not hereditary. But meanwhile evolutionism itself evolved in the course of the twentieth century. It was discovered that evolution does not follow an unbroken line as has been supposed, but that it jumps in a number of places. This theory of "mutations" had already been proposed by Thomas Henry Huxley (1825–1895). In its modern attire it teaches that changes take place in the molecules of the chromosomes, the DNA. This very long molecular chain composed of so-called neuclotides bears the properties of heredity. A nucliotide is thus a building-block of the DNA and it is composed of an organic alkalid, a sugar, and a phosphate. The major changes in the chromosomes are called mega-mutations.

Professor Søren Luvtrop of Sweden established that especially the mega-mutations are responsible for the rise of the new forms (the new species in the animate world). An eventual confirmation of his theory would imply that the frantic search by paleontologists for the so-called missing links (forms supposed to fill the gaps that break the presumed continuity of evolution) would all be in vain. Nature would simply have skipped them. I have to admit that I once indulged myself in this sort of fantasy during my works on determining the fossils of large collections from Egypt and Sinai. I amused myself at arranging them in long rows, various bivalves of the *Cretaceous Exogyra* or of the *Eocene Ostrea Appendicularis* and I gave my imagination full reign how one form may have developed from the other. I am aware that this kind of game may hardly be called being engaged in science.

Much more interesting are the discoveries from experimental research. The results have been beautifully summarized and revealed to everyone in a paper by Robert F. Weaver in 1984, "Beyond Supermouse: Changing Life's Blueprint," *National Geographic*. The technique of changing heredity in the laboratory is called "gene-cloning" of the DNA. "In effect, gene-cloning is like cutting a printed page (he compares DNA with the printed page) in half, inserting a new paragraph in the middle and photocopying the altered version over and over to reproduce the new material along with the old." We may of course doubt if nature ever produces its new forms this way. We are not even sure if the monstrosities we turn out are viable, and I believe that it is equally doubtful if we may ever be able to witness the emergence of a real new species in nature (not a thing like congenital mongolism) as the very act of our perceiving the event may disturb its natural course, and not to mention the very slim chance that we may be present at such a rare event.

After it has become clear that we had to reject the original form of Darwin's thesis, and after one began to understand the obvious, that the progeny is always born as silly as its parents had been before they acquired new tricks, one began to focus on the real problems. Take the following questions for example: Why do macro-mutations never take place along haphazard and arbitrary lines? How it is possible that the changes appear to follow a direction that benefits the species in its adaptation to the new circumstances? This serious problem is connected with Immanuel Kant's philosophy on purposes of organs and organisms. I recall that we concluded that there is no question of a willful adaptation because plants and animals do not experience the changes consciously.

Another serious question is whether there have not ever been any mishaps. Or may it be that paleontologists are unable to detect them because fossils are stone-dead and never wrote their autobiographies? Anyhow, we do not know anything about the statistical relation between misses and lucky strikes in evolutionism. It, however, seems if a leading hand guides evolution; one may be tempted to write the leading hand with a capital "H." For if no living creature improves itself through its own free will, whose will may it be? Let us analyze if, for example, the leading hand that takes care of progress in the animate world may be adduced as a new form of argument to prove God's existence. A special form of argument from design, a close scrutiny, must disappoint us. The leading hand that is supposed to convince us of God's existence deals out its evidence in support of this argument from design in such measly, scanty measures and installments that we may hardly talk about progress. It goes at a snail's pace and it would be a blasphemy to assuage this progress directly with God's name.

I propose therefore to stop looking at evolution with the eyes of a Cartesian dualist and to regard all these interpretations as one-sided. Organs and organisms are only the psychical aspect, interpretations from our sense perceptions. Under the aspect of mental reality the problem will turn out less of a mystery. Let us regard the mental world as a complex of—in the main unconscious—mental units in which only a very special structure leads to the conscious. We should however be aware that we are not able to penetrate into the real character of this mental reality became our senses are the only poor means to gather information about this reality.

It seems to me that the leading hand is something mental acting in a mental world; in other words the leading hand is part and parcel of Creation and not God's own "Hand." It is more than some artificial limb created by God, fighting, in a rather clumsy way, in the struggle for survival on behalf of God's creatures, who themselves are unaware of what is going on. Charles Darwin may have been well aware of this. The famous slogan "survival of the fittest," was not his invention but that of Herbert Spencer (1820–1903) who turned evolutionism into a philosophy. Henry D. Aiken remarked in his 1956 work *The Age of Ideology* that "Darwin did not intend a teleological theory that imputes to all organic beings a purpose to live." This was the naïve idea of Thomas Aquinas and of Immanuel Kant.

What was the reaction to Darwin's works at the time he published? The early Darwinists and the clergy promptly opened a silly tussle.

Both camps laid Darwin's evolutionism alongside the Bible—the valid truth on the one side and the moral value on the other. But both ignored the very essence of the Bible that is meant as a moral message. The Darwinists tried to prove that the Bible was nonsense and the clergy tried to show that their opponents were committing the most sinful heresy. The two sides could as well have laid a textbook of zoology alongside the fables of La Fontaine and then have ignored that their real message is a moral one. Only children fail to understand this. They look high upward in the trees to catch a glimpse of Maître Corbeau sitting on a branch with a piece of cheese in his beak. But both the Darwinists and the clergy behaved likewise. They both studied the Bible to the letter.

The Archbishop of Canterbury retorted that if man developed from the animals whereas man has been endowed with a soul, then even the worms must have a soul. I am ready to confirm that worms do not have souls—they *are* souls, though unconscious ones—and believe it or not this kind of silly war is still going on. The general commotion began to increase with the appearance of Thomas Huxley's work in 1863, *Evidence as to Man's Place in Nature*, containing his discovery that the anatomic differences between man and the anthropoid apes are of minor importance compared with the gap that separates the apes from the build of the lower monkeys. This was seen as a direct challenge to the clergy. It seemed to undermine the unique position of man above the animal world, the very foundation of theology. This extra insult came on top of the challenge to Genesis. If we allow the clergy to stick to their guns, permitting them to compare Darwinism with the Bible, we would still have a counterargument up our sleeves.

Darwinist evolutionism does not challenge Genesis, but evolution has simply been omitted from the Bible. It is written in such a manner and fashion that simple people would promptly understand it. The Bible was meant for simple people to understand and it therefore recounts in Genesis 2:7: "And the lord God formed man from the dust of the ground, and breathed into his nostrils the breath of life." And I am well aware that this conception is suggestive of a dualistic Creation, being composed of "dust of the ground" and of "breath of life." An account compatible with our "monastic" point of view would however, be totally incomprehensible to the common reader.

And even if the clergy would persevere in misreading the real purport of the Bible by comparing Darwin and Huxley's descent of man with the variation in the Bible, we may explain that the dust of the ground may mean nothing but the world of space and matter, compromising both inanimate dust and the physical aspect of the animate world. We may then finally comment that it does not make any theological difference if man would have been created directly from dust of the ground or through the long chain of lower animate forms.

But I certainly maintain that we had better dispense with all these devious twists and counter straight away with: "Look at the message of the Bible and be satisfied." But the Church of England regarded Charles Darwin as a defective apostate, a dropout from his creed who betrayed his original vocation to become a priest. I refer here to the 1962 work by Gertrude Himmelfarb, *Darwin and the Darwinian Revolution*— a biographical, historical, and philosophical study of the impact of Darwin on the intellectual climate of the nineteenth century. "Was it not more natural to conceive of the soul being breathed into the body of man, rather than invoke a double miracle by which the animal soul was first breathed out of him and the human soul then breathed in?" Swiss Protestants would indeed read in this another reason to disapprove.

Ulrich Zwingli posited the opposite: that man was supposed to have been brought into being by his soul being breathed in as God's Holy Spirit, but that Adam lost it and that he and his progeny have to bear the burden of an animal soul because of his fall, to bear a "prast," as Zwingli called it. "This means an illness far worse than to be labor-stricken." Zwingli meant that through the pains of Adam's wife and the ultimate death of both, man becomes an animal. It is however not my task here to deviate between two untenable positions of the Protestants.

Let me add some words of comfort to all the stubborn members of the clergy who fear that Darwin and Huxley tried to undermine the position of man which they all believe to have been safeguarded by Genesis 1:25–28, which means by God ordering man created in His image to populate the earth and to rule over it. I read a compromise in Ecclesiastes 3:19: "...that a man has no preeminence above beast," though said in the context "that which befalleth the sons of men befalleth beasts." We may also read in it that "which befalleth beasts may befall man" as a warning that man may become beast if he does not obey the moral laws that safeguard his hegemony of the earth. What else may be the reason why one has to repeat this sentence from the Bible several times on the Day of Atonement, the day of moral reckoning for the Jews?

And let us now turn to the other camp, the defenders of classical Darwinism, Professor Stephen Jay Gould of Harvard among them. Even

Darwin's own friend and pupil Alfred Russell Wallace (1821–1913) who put up a theory all of his own conferring in detail with Darwin's evolutionism, was in Gould's eyes a dangerous heterodox. Why? Because he dared to disagree with Darwin on just one point. Wallace remarked that many phenomena developed in the human spirit that do not tally with Darwin's because they are of a special nature, completely unknown in the animal world. Gould revolted against this heresy and retorted that "many serious researchers put a picket fence with a sign on it around their own species." This is an intolerable accusation leveled against Wallace's serious argumentation.

Wallace could not be accused of an anthropomorphic bias. His arguments were based on a clear matter of fact. Man developed special abilities that could only impede and certainly not promote the survival of the fittest. Among them he counts artistic, scientific, and moral faculties. It is indeed beyond doubt that artistic proclivities are not to be counted among the vital conditions to keep us alive, as an American musicologist once said, "music is not meant to make a living, but it certainly assures a better life."

The inner urge of a scientist is an expression of pure inquisitiveness and mostly not the pursuit of a practical purpose, while we are all convinced that moral virtue is a property that stands far above the instinct of collective self-preservation. Wallace rendered these opinions in the last chapters of his work, *Darwinism, an Exposition of the Theory of Natural Selection*. In the book he quoted a certain Dr. Weismann saying on heredity: "These predispositions—which we call talents—cannot have arisen through the natural selection because life is in no way dependent on their presence." Further, Weismann remarked that this "clearly points to the existence in man of something we may refer to as being of a special essence," and he concludes, "thus alone we can understand the constancy of the martyr, the unselfishness of the philanthropist, the devotion of the patriot, the enthusiasm of the artist, the resolute and persevering search of the scientific worker after nature's secrets." This is certainly not something "which he has derived from his animal progenitors."

Wallace accepted this but he felt that it implies a problem: how is it possible that man may rise above the laws of nature that define the developments of the animate world? He asked a three-fold question: (1) how did the animate rise from the inanimate; (2) how could the animate world acquire its generating power to propagate; and (3) how does life create consciousness?

From Wallace's questions, it emerged that he had emancipated himself from Darwin's shortsighted materialism. On the other hand, he did not manage to free himself from Descartes's dualism. When we reduce all these processes to the category of mental reality we may reformulate Wallace's questions as follows: how could human consciousness arise from the animal unconsciousness? Herewith we have eliminated the fictitious leap from the physical to the mental. What remains to be explained is how human consciousness could arise from animal unconscious.

We observe indeed that animal behavior follows a fixed pattern and there is probably not one animal which is aware of its behavior. This means that the gap between the human conscious and the animal unconscious is much wider than the discovery by Thomas Huxley would suggest. That the gap between the anatomy of the anthropoid ape and of man is rather narrow, the gap is in reality a qualitative mental one our senses fail to observe. Darwin the materialist tried to blur this gap. He regarded a man as just the last link in the unbroken chain of evolution. But Wallace knew better. Professor Stephen Gould on the other hand tried to defend his co-materialist Darwin with all his heart.

We end with a piece of evidence that the tug-of-war is still in progress. At least at this moment of my account. The two camps, the Darwinist and the Jewish orthodoxy, convened in March 1983 in Jerusalem for the first congress on inquiries into the origin of life and evolution. Once again the Bible, this time the Old Testament, was laid alongside the classical works of Darwin and Huxley. The two sides behaved as if they had been invited for a bottle-party where only guests are admitted who take the bigotry of the personal inhibition with them. The discussions were laden with emotional arguing. Several "scientists" had obviously their own axes to grind and dispensed with scientific objectivity, and the other side behaved as judges at a religious synod. Sincere efforts to introduce some reason were not wanting.

Dr. Patrick Frank of the Weizmann Institute in Rehovoth, Israel, published a letter in the *Jerusalem Post* of 13 June 1983. "A sound theory about the origin of man has to explain how an antibody specially composed to react on an albumen of a certain species also reacts on similar albumina of closely related animals and that the intensity of the reaction gradually decreases in creatures which differ more and more from these animals." It turns out this decrease follows the supposed line of evolution deduced from paleontological research. Frank concluded his letter with an inference that all these variations in composition of the albumen in plants and animals, such as hemoglobin, cytochrine, immunoglobin, clearly point to a common ancestral carrier of a "primeval" hemoglobin and other primeval albumina. And this, so he says, is the clinching vindication of evolutionism.

Frank is of course justified to table all this material as supporting evidence and I do not underrate the importance. Supporting evidence, however, is not yet an irrefutable proof. He only demonstrated that in addition to Darwin and Huxley's anatomical similitudes on which the original evolutionism has been based we have to take the newly discovered biochemical similitudes into account. We all feel that this parallelism strengthens the arguments in favor of the theory of evolution.

I have to add a few words of caution: the use of the term "relationship" alongside "similitude" may be misleading. "Relationship" suggests genetic affinity, which is not at all the same as resemblance. Mineralogists dare to speak of the family of the feldspars, but they would never suggest that this term must imply that one kind of feldspar would arise from the other. In the same way chemists are well aware that a term like "the family of the halogens" may not imply that chlorine originated bromine.

My final moral can only be a moral about morals. I compared the futile act of comparing Darwin's evolutionism with the contents of the Bible, with laying a text book of zoology along the fables of Jean La Fontaine (1621–1695). We have an unshaken faith in the contents of a good textbook on science. But do we believe in fables? No. But if we would ask if we believe in fables as a means to illustrate a moral value, ours would be an unqualified yes. Let us compare this answer with the act of laying evolutionism alongside the Bible. The books of Darwin and Huxley and Wallace do not convey an absolute scientific truth. They were three giants in a collection and compilation of unknown facts, but there is something amiss in their final conclusions.

No textbook will ever convey the absolute truth. The developments of science means science in progress—something ever-improving is never ideal. One has a right to regard the Bible as unassailable only under one condition, as long as one regards it as a means to illustrate an unassailable moral value. As soon as we fail to regard the Bible as a message and misread the illustration as reality we are in trouble. I remember a Dutch priest who, during the period between the two great wars, also took his colleagues under fire on the issue of whether the serpent really talked or not. World War II brought this discussion to a sudden end. I told you that the children who listened to the fables of La Fontaine may behave likewise.

The Bible is a means to illustrate moral values, and I regard the absoluteness of moral obligation to be so far from self-explanatory that it is only intelligible to me in a theistic context. On this point I am in complete agreement with Illtyd Trethowan, who in 1960 published *The Basis of Belief*.[1]

Karl Marx and Charles Darwin stood at the nadir-point of the downward trend of religious devotion. Their atheistic attitude can only be matched by the atheism of Bertrand Russell who came a century later. Marx had been the cause of the greatest social upheaval in the nineteenth and twentieth centuries. The consequence of Darwin's evolutionism was much more modest. He was after scientific truth and never promoted social or political views. But the combined effect on nineteenth-century religion and theology had been severe. Christendom had a hard time to defend itself, weakened as it had been by the schism a few centuries before. It is a matter of course that the Catholics and the Protestants reacted differently. Catholic dogmatism has been able to fossilize the mother church to a degree that it could not be undermined from within. Protestantism had always been a bit more amenable to change that signified change, a change of rituals. Some of its vicissitudes can only be explained as a hasty putting up of a theological barricade against the rising tide of unbelief.

And the Jews—this poor people had to adapt itself to all these evils combined. If I would follow the same course as in my account on the Middle Ages and start with the development of Judaism since the Reformation, my account would look like a map without contours, as a haphazard pattern of meandering roads, and the twists would not be intelligible as following the easiest routes to avoid difficulties. I decided, therefore, to take this time a different course. The history of Christianity comes first and this will make the twisting curves of the development of Judaism more understandable.

Note

1. *The Basis of Belief* was quoted by Brian Davies in *An Introduction to the Philosophy of Religion*.

14

An Introduction to Ignorance

What God May or May Not Have in Store for Us

Futurology is the systematic forecasting of the future in general by extrapolation from the course of the processes which brought about the known present situation from the known past situation.

The past and present may conceal
Whatever the future will reveal

After God has molded the universe, His part in molding the future is decisive. From our human vantage-point, the future has many aspects. First of all I may explore the various past opinions about the future generations. In these pre-scientific times people tried to read the future from the Bible, and the study of these predictions is called "religious eschatology"—derived from the Greek *eschatos* meaning "the ultimate."

Quite another way of looking at God's part in molding the future is the modern scientific approach. What may be the future of the earth and of the animate world? What may we guess about the future of the universe as we may derive from its present state, provided that we know enough about its present state? Thirdly, we are interested to know as much as possible about our personal future, and I mean here about our eventual afterlife. This problem is part of the discipline called *thanatology*, the scientific study of natural phenomena related to death.

But I believe that the most important question is how we can change the world for the better, and how we are expected to perform in order to bring the improvement about.

As to the first problem, the history of the opinions among theologians of the past generations—in short religious eschatology, we are on solid ground. It is a subject that has been studied in detail, and consequently, we are enriched with a vast amount of literature, much of it of superior quality. I recommend the excellent compendium by Norman Cohen, *The Pursuit of the Millennium*. Though my paragraph on this subject is the most elaborate of part 3, I beg you not to overrate its reliability as a source to read the actual future. The methods used by the theologians of the past did not bring us one step closer to the truth.

As to the second approach—what God has in store for the earth, for the universe—I am not able to satisfy you with any definite answer. I can do no more than furnish you with a summary of some different viewpoints.

What We Are Thinking about on Every Subsequent First Day of the Rest of Our Lives

This section does not contain more than the raw reflections of the person-in-the-street who does not take the trouble to separate the logical conclusions from their emotions. I can hardly call it "science."

We concluded earlier that time is a real parameter not only in the world of our senses or physical world but also in the world of mental reality. I recall that Roger Penrose concluded that time in the physical world must be irreversible because of its special relation to gravity, and from this emergence we have to infer that time must be irreversible in the world of the mental reality as well.

We have convinced ourselves furthermore, from mere logical reasoning and from modern discoveries, that time has been created and hence that the world has been created. This past event looks very distant if we confront it with the duration of our own lives. The certainty that the world has been created seems to have little to do with our certainty that our own life had a beginning. The first certainty is the certainty of a flawless logical deduction, and the other certainty, that our own past is finite, follows from our reliance on witnesses, such as our parents, and from our own experience that tells us that people are born.

Our conviction that the duration of our future life is finite too, is equally the fruit of our experience, though a slightly different one. We understand that potential infinity is a reasonable possibility, but we are aware that this possibility cannot be applied on our life. When we look ahead we must conclude that the statistics are not inspiring. It is however a very unhealthy habit—and now I am speaking with the experienced psychologists—to keep moping about our inevitable fate. I know that there are a great many, and they are not only existentialists, who borrowed this morbid pessimism from Christian theology, and who, as a consequence, abandon themselves to a listless condition because of their preoccupation with a grotesque misinterpretation of life. They see themselves as having been put quite beyond their will, on some conveyer belt at the moment of their birth. Their face is turned towards the trailing end, which is the past disappearing out of sight. They feel that their back is turned to the leading end, the future, because they are not endowed with eyes in their backs nor are they able to turn their heads.

They do not know when and where they must be, inevitably, bumped off from the conveyor belt of life. They are, in short, not very different from the chained audience of Plato's den, imprisoned for life to be executed on an unknown day in the future. They are the symbol of helplessness and hopelessness. This is not life but malicious characterizing. It is a view which kills every notion that life must have value. Puppets, even puppets endowed with consciousness, are devoid of this instinctive notion of value.

We are aware that the most valuable among values is moral value. But we are unable to define the value of life. We are not even able to give a rough description. Long ago I overheard a statement that "the value of life is being alive." It sounded to me as a cryptic pleonism, as a Kantian analytic statement, as a tautology like "the value of money is money." When I grew older I came to realize that there is more behind the terse sentence. I found it out the hard way.

We are not aware of the value of life before the moment when someone for whom we care, and who cared for us, is taken away. Then we feel, all of a sudden, the value of life, of our own and of others. It is the nagging pain of bereavement. Normal people have an inherent urge to live on and on. Let me recall that we encountered this urge in the chapter about Spinoza's philosophy where he called it *conatus*. The reason why people do not want to die depends on their character. Good people do not crave for it for the same reason as bad people. A bad man may have overcome mortal fear, but he still begrudges the glee of his enemies who survive at the moment of his death.

The arguments of good people are just the reverse. They loathe the prospect that the circles of their surrounding friends would suffer the same pain that they must have experienced themselves during their life at similar moments. The aggravating circumstances is that in so many cases this circle widens and widens. More and more people are added to their circle. First a wife or a husband, then the children then the grandchildren. Good people have to suppose that all the members of their circles entertain feelings which are similar to their own. They surmise that their friends are equally dear to themselves and for the same reason. In short, good people respect all these supposed feelings in others, which they detect within themselves. They have, in fact, this respect for everyone whom they happen to encounter. At the end this universal respect has to be identified with the principle of objectivity as defined by Gerard Heymans.

May all the physicians of the world make a mental note of these edifying motives and may they all join hands in their efforts to promote our physical and mental health and prolong our lives. Mental health is just as vital for the patient as for his domestic circle. A long life and a deteriorating mind are a curse. A father who does not recognize his son anymore may not suffer, but it is worse than a burden for his son, it is a source of agony.

Let us compare this natural and healthy loving care with the attributes among so many heirs of Christian theology; among the mystics who contemplate some form of mental self-mortification, the ascetic dreamers of slow self-mortification, which they regard as the ultimate act of piety for the sake of an alternative they mistake for being the "better" in the great beyond.

They are convinced that this world comes to an imminent end, and that all their doings in the world are meaningless. The prototype is the monk and the Catholic priest. Their first act was isolating themselves from their family and the world and with this sinful act of running away from their moral obligations, they claim to prepare themselves for the salvations of their soul. Small wonder that a great many among them feel the frustration of committing a mortal sin in following what is a sinful commandment. They are utterly helpless and at a loss how to overcome their doubts in the rightfulness of their desertion. They have put themselves—in their imagination—amidst all the others who are sitting passively on that conveyer belt, and obsessed by the prospect of their ultimate deaths, they cannot think anymore about anything else.

Existentialists do not behave better. Their basic principle is that philosophy is obliged to concern itself with the human predicament and inner states, such as alienation, anxiety, inauthenticity, dread, sense of nothingness, anticipation of death.[1] Life has been emptied of everything that is of interest for a normal being, with the exception of an obsessive interest in the anticipation of death. They even detest their own birth.

What is the fatal error of all these doomsday believers and all the Christian futurists? That they misinterpreted the obvious facts that they are born and that they are bound to die beyond their will. Both events have been decided and will be decided by factors in the external world. This refers not only to our existence, it refers also to our essence. We are what we are. The pot is not allowed to blame the potter for the deficiencies in its essence, not even if it would be permitted to write "potter" with the uppercase "P" and point to our parents. We have not the right to question the rights of our parents to bring us into being, nor are we allowed to judge if it is righteous that man is mortal. Birth and death are two points in our lives that are none of our business. What matters is how we fulfill our obligations between these two moments.

To be created is to be alive. Life and living entail problems that we cannot solve by kneeling for God's so-called omnipotence. We serve him only if we act, and when moral issues are at stake we have to behave accordingly by our acts of loving care. Life does not know of higher ends. We have to show our loving care not only to the individual but to the entire world of creation. We have to muster all our strength and make use of all our limited faculties to maintain and to improve the world, especially from the moral point of view. We have to show that we are worthy of having been created.

All this entails the obligations that we should never believe in an imminent, not even in a future, perdition of the world. We shall see that science does not warrant a doomsday religion. A doomsday religion is not only silly, it is sinful. It holds God's creation up to scorn and ends in holding God up to scorn. Shun this blasphemy!

No less sinful is a recognition that does nothing but stress and over-stress our nothingness. God did not create "nothing." If your are so convinced that your are "nothing," try and make something better of yourself; challenge God if you dare. What I have in mind is a tremendous task, even if all the people of the world join hands. If you don't feel equal to your task you may pray for a strong back. This is the kind of prayer that does not only edify you, such as thanksgivings, it may mortify you; it may help to impart vigor and mental strength. But this effect is not a purpose in itself, just as the loading of a battery is not a purpose in itself. The enhanced energy has to be put to good use.

These lines convey the thoughts of the man in the street; they are not based on any scientific, philosophical deductions. The subject does not lend itself to this standard of study, but we may at least look into some linguistic aspects.

The word series—"sense," "meaning," "value," "quality"—does not leave much room for sharp distinctions; their verbal significations merge one into the other. "Value" and "quality" are judged by value and quality scales. We speak of higher and lower values and of good and bad qualities. "Meaning" has its grades too; a life may be more or less meaningful. One of the differences between "meaning" and "sense" is that meaning is more personal. Something may have no meaning for me, but to state something about sense of life is uttering something that has a tinge of objectivity. "Sense" is a different word in relation to life. If we ask of the essence of life or of the use of life—"use" not to be associated with "useful"—we ask in fact, "why life?" And to God we ask, "Why did You, God, create life?"

But though we have to leave this question out, though we will never know anything about the sense of life, and do not even know what it signifies, we may at least expect from ourselves that we live a meaningful life, and that we feel that meaningful is associated with virtuous. Though we are all convinced of the truth that moral virtue makes a life meaningful, it is obvious that the emotional character of terms "meaningful" and "virtuous" preclude a successful proof of this statement.

Note

1. See Peter A. Angeles, *Dictionary of Philosophy.*

15

Religious Eschatology,
Part 1: The Jewish Messiah

The Jewish Messiah made its appearance rather late in the records of the Old Testament. The short account that follows here has not been written in the vein of a faithful believer, but is of an objective historian who tries to abide by the facts.

The Holy Land, where the Jews ordered their life, was most of the time under the influence of, and at least under the indirect rule of, foreign empires. The Bible tells us remarkably little about this form of tutelage by the superpowers, but as soon as they tried to interfere with the Jewish way of life, especially with the religious aspect, their vile acts became the central subject of religious interpretations, such as heavenly punishments on account of moral iniquities of a Jewish judge, the depravity of a king, of the remissness by the people in the fulfillment of religion obligations.

Alexander the Great subjected the Holy Land in 332 B.C. after he managed to wrest it from the Mesopotamians without serious opposition. He did not interfere with the life of the Jews, and the Old Testament practically ignores him.

Alexander died in 323 B.C. and from that moment onward things went awry. His surviving generals, the "*Diadochs,*" which means "successors," began to quarrel among themselves over the imperial inheritance. The empire was split up into Hellenistic states. Egypt was ruled by the Ptolemies, Syria by the Seleucides, Macedonia by the son of Antigonus Gonates and all these altercating kingdoms fell prey to the common enemy, the expanding Roman Empire.

Ptolemy I of Egypt succeeded in putting his rivals aside but a few years later he had to stave off the attacks of the Seleucides. The Ptolemaic rule came to an end, the Holy Land fell into the hands on Antiochus II (223–187) who managed to keep the country (in the year 200) more or less under control. "More or less" means that the Ptolemic influence remained very much in evidence even until the final conquest by the Romans.

It seems that the Jews have suffered all these superpowers with a fair dose of equanimity because they did not change their everyday life very much. But no sooner did a Greek king try to impose his Greek way of life on the Jews that all hell broke loose.

Antiochus III suffered a defeat by the Romans and fell into debt. None among all these kinglets paid much attention to tiny a Jewish people, which was either ignored or oppressed. The Levant was only a much coveted link between the rising cultures of Greece and Rome and the declining cultures of Mesopotamia and Egypt. How could a Ptolemy and Antiochus guess that all these civilizations would disappear within a few centuries and that the Hebrews would engender religions that would encompass the entire world?

There was, however, in these days and in this region one conspicuous manmade landmark adorned with outstanding valuables, the Temple in Jerusalem. It was too tempting a source of capital levy to be ignored. Seleucus IV decided to square his debt by taking the equivalent of his deficit from the riches of the Temple, and his brother Antiochus IV Epiphanes (175–164) committed the additional folly to depose the high priest, Onias III, and to replace him by the Greek traitor, Jason. What happened thereafter has been eloquently described by professor Menachem Stern in his contribution to the Keter anthology, *History*. The chapter is titled "Second Temple: The Hellenistic-Roman Period 332 B.C.E.–70 C.E."

Stern wrote:

> The unlimited devotion of the Jewish masses to their religion was in any event deep-rooted but on this occasion there unfolded, for the first time in the history of mankind, an epic chapter of martyrdom on a large scale. To bequeath in the resistance of the martyrs and the Hassideans (religious rebels against the persecutions) a symbol and an example throughout all succeeding generations to both Jews and non-Jews.

It was during these insurrections that the messianic expectation was born. I believe that this episode was so crucial in religious history that we may say today, with confidence, that without this so-called "Hasmonaic Revolt" Judaism would not have survived the Greek efforts to Hellenize the Jewish people and that consequently Christianity, Christianity that emerged, by the way, as the arch enemy of Judaism, would never have been born.

We may guess that verses 7–8 of Jeremiah's prophecy, 23, had been a source of inspiration. "Behold the days come, saith the Lord, that they shall no more say, the Lord liveth, which brought up the children of Israel out of the land of Egypt; but the Lord liveth, which brought up and which led the seed of the house of Israel out of the north country and from all the countries whither I have driven them; and they shall dwell in their own land."

These lines became, by the way, also a source of inspiration of the Jewish people thousands of years later when it founded the Zionist World Movement and when its program was fulfilled with the independence of Israel. The Lord pronounced in this prophecy a bold sentence that provoked some serious questions among the ancient sages. One of the principle commandments in Judaism is the yearly commemoration of Exodus on the seder evening or the Passover evening—the miraculous departure of the people of Israel from Egypt. Did God seriously contemplate overruling His own commandment? Because He would perform a still greater miracle, the gathering of the Jews from the Diaspora, a calamity that began with the Greek oppression. We find the reference in the last Mishna of the Talmud, "Berachot," in a discussion between Ben Zoma and other sages. The answer was that one is obliged to keep the commandment also in future, the Seder evening should never be abolished. Future events would never overrule but overshadow the memorial evening consecrated to Exodus. The delivery from foreign oppression is the decisive point (see Berachot 12:63) and the Gemora says that the only difference between the Messianic times and the present (the "present" means the early Middle Ages) would be the absence of foreign rule. The essence of the Messiah is freedom from slavery. Maimonides added that one should never believe in the abolishment of a law of nature. The world, God's creation, would not change; only the human spirit would be revolutionized.

Though Maimonides did not deny that the coming of the Messiah would be decided by God, the event would have, in principle, a worldly effect. We should not scorn this interpretation as a profane secularization of the Messiah. Maimonides was convinced that our everyday occupations would be hallowed through good works to be performed in a spirit that precludes the prospect of any reward.

Herman Cohen, who wrote *The Jews and Judaism after the Reformation,* wrote in 1907: "It is the duty of any Jew to help bring about the messianic age by involving himself in the national life of his country."[1] Though the meaning of "his country" is a matter at issue between Cohen and me, I admit that we should believe that the coming of the Messiah depends on our behavior, which implies "involving ourselves in the national life."

But the views of Maimonides and Hermann Cohen may nevertheless lead to profanation of the Messiah if we go too far—Martin Buber's rule of life should be rejected. We find Buber's views in Gershom Scholem's *Martin Buber's Conception of Judaism*. Scholem was one of the very few who was able to cull some sense from Buber's obscurantism. Buber seemed to believe that the Messiah is being realized in every single positive act, as if it were as an early counterpart of heavenly creation which Buber—erroneously—sees as a continuous divine process. We may value the truth of this statement though introspection. I may fancy that I would be edified through some of my profane acts and I am ready to believe that such acts, by multiplication, might bring the coming of the Messiah closer. But I believe that it would be far-fetched to maintain that divine redemption realizes itself within us through these trivial acts. I do not decry the holiness of good works but it is clear that Buber's metaphysics have nothing in common with the current views on the Messiah as they are represented in the Talmud. Let us therefore not go farther than Hermann Cohen's statement that "these acts may help to bring about the messianic age."

Let me however return to the concluding words in Hermann Cohen's sentence, "involving himself in the national life of his country." I recall that "his country" meant for Hermann Cohen every country where a Jew happens to live, and that this interpretation had been exactly the cause of the final doom of European Jewry a century later. It is obvious that a Jew should involve his national life in just one country, the Holy Land, and if he happens to live elsewhere he should make it at least his object to live in the Holy Land in order to make the sincerity of his involvement in the national life of his country more credible; and if he happens to be religious he should be aware that this attitude is not just common sense, but the fulfillment of a divine commandment as follows from the closing verses of nearly all the books of the prophets in the Old Testament. The building up of the Holy Land, which implies its spiritual edification, is like the bricklaying of one stone upon the other, and do not forget the mortar in-between.

Let us go back to the Bible for further documentation: Ezekiel 37:12 says: "—thus saith the Lord God behold O My people I will open your graves and cause you to come out of your graves and bring you into the Land of Israel." The extermination camps in Europe were indeed the "valley full of bones" (Ezekiel 37:1) and the survivors could not feel any other urge but to be brought into the land of Israel.

Joel 3:20:21 says: "But Judah shall dwell forever, and Jerusalem from generation to generation. For I will cleanse their blood that I have cleansed for the Lord dwelleth in Zion." Amos 9:14 says: "and I will bring again the captivity of Israel and they shall build the waste cities and inhabit them...."

Going back again to the discussion between Muki Zur and Gershom Scholem[2] and starting at the point where Zur reminds Scholem that he once said that Zionism is not a messianic movement, whereupon Gershom Scholem answered that the Zionist movement would have been doomed to fail if it had pronounced itself as a messianic movement. He reminded us, of course, of earlier movements like the one headed by Shabbatai Zevi in the seventeenth century. This is a wise admonition, but it is, on the other hand, impossible to dissociate in our mind the Diaspora and the return of the Jews to the Holy Land from the messianic idea, and if we rarely obliged to heed Gershom Scholem's warning, we should regard Zionism as one of the main tools to be used in our good works that may help to bring about the messianic age. And we are not able to aim higher; Zionism is a human and fallible effort.

It is remarked, though not always realized, that practically none among the prophecies in the Old Testament point directly to a personal Messiah, though Ezekiel 37:25 explicitly states "and my servant David shall be their prince forever." The name Messiah does not appear at all but much later in the New Testament in St. John 1:41: "We have found the Messiah, which is being interpreted, the Christ (the anointed)" and in the same gospel, 4:25, "The women saith unto him I know that the Messiah cometh which is called Christ." The Hebrew root-word "*nasach*," or "anoint," from which "the Messiah" has been derived, appears in psalm 2:6 "*ve 'ani nasachti malki al-Tsion har kodshi*," or "And I anointed you as my king upon my holy hill of Zion." Though this verse is suggestive it is not a matter of course that it has to be associated with the above quotations from the prophets. It was not the Bible, but Jewish folklore that turned David, King of Israel, into the living Messiah. In Jewish philosophy, "Messiah" had more abstract meaning, the ultimate element of a religious triplicity that returns, in a somewhat different form, in Christian theology—creation, revelation, redemption among the Jews (as defined by Martin Buber and F. Rosenzweig) and as Creation, Final Judgment, and Redemption among the Christians.

"Revelation" in Judaism is not the revelation of creation, nor the divine prophecy of redemption, but the revelation of our Creator and Redeemer in the theophany on Mount Sinai where He appeared through His act of the *Mathan Torah,* His delivery of our moral beliefs in the verbatim form of the Torah. That these moral beliefs existed in the human soul before the *Mathan Torah* is obvious. The Jewish people would not have been accessible to the contents of the Torah nor would it have been ready to accept it if it would not have been endowed with moral sense, which mankind received through the tree of the knowledge of good and evil. This connection refers, however, to the universal moral sense of the individual, but how was justice administered to the Jewish people before it received the Torah? Yeshayahu Leibowitz[3] pointed to Exodus 18:13–27. Moses camping in the wilderness of Sinai judges all his people as the one and only magistrate and Jethro asked "why sittest thou thyself alone and all the people stand by thee from morning unto evening?" And Moses answered his father-in-law *"Kie yavo elai ha'em lidrosh Elohim*—because the people come unto me to inquire of God." And this may mean to give homiletic interpretation of God as a moralizing discourse for spiritual edification, but summarized, of course, with a judgment, with passing sentence.

The direct connection of human moral sense and God, the Creator of moral sense in the human soul, was a matter of course for the children of Israel and for their judge, Moses. To inquire of God and to be judged by Moses are one and the same and all this happened before the episode of Mount Sinai. The people were judged according to *ius naturale* or according to Gerard Heymans's principle of objectivity, which is just the same. "I do make them know the statutes of God and His laws," said Moses (Exodus 18:16). I regard this as Moses' admission that the intuitive *ius naturale* lives in everyone, but that it had still to be formulated as normative *ius naturale,* and he lived in eager expectation of what was to come in God's delivery of the "Book of Enlightenment"— the Torah.

The other words in the Bible are self-evident. Moses could not continue to fulfill his task all of his own or as Jethro said, "Thou are not able to perform it thyself alone." Consequently Jethro gave counsel that Moses should appoint an executive body, a governing administration. The persons to be elected, his future staff of judges and magistrates, should be endowed with the moral virtue that raises them above the judged. Once again moral virtue predated Mount Sinai and was there since the creation of man. Hence moral virtue will guide him as long as he will live on earth, before the Messiah and after the Messiah, and the Torah will always enlighten him. The Messiah will appear within a finite time and God will not release His creatures from their moral obligation.

Though this potential eternity of the validity of the Torah is not explicitly formulated in the text of the Old Testament, we find a clear reference in the weekly prayer following the reading of the Torah on Sabbath morning— *"Baruch ata Adonai Elohenu Melech ha'olam asher nathan-lanu Torat emet ve'chay'e olam nathan betochenu*—praised art though O Lord, King of the world who has given us a Torah of truth and planted eternal life in our midst." Eternal life does not refer to the eternity of our soul, but to the potentially infinite future of the Torah, which shall exist as long as mankind shall exist.

The prayer on Sabbath morning is in fact a sentence from the Talmud, Sopherim 13:8. Indirect hints in the Old Testament that God's commandments are binding forever are numerous. Exodus 31:16 is another example: "The children of Israel are commanded to keep the Sabbath, to observe the Sabbath throughout the generations for a permanent covenant." And Exodus 29:28 speaks of a "statute forever from the children of Israel." Skeptics sometimes challenge me to point to a written obligation in the Old Testament that the Torah is eternal. But its serenity is self-evident and the Bible does not prove anything it introduces. Genesis introduces God with the front page of His identity book:

Name: God

Age: Eternal (not mentioned in opening lines of the Bible)

Address: Ubiquitous (His Spirit was known to have moved upon the face of the waters)

Genesis then gives a short curriculum vitae detailing God's achievements as Creator of the world. It is just one chapter of thirty-one verses. If the Old Testament does not intend to give evidence of God's eternity, past or future, why should it give evidence of the eternal validity of the Torah?

If the Torah will be binding and not lose one jot or tittle of its validity after the coming of the Messiah, will it also be a binding obligation to rebuild the Temple in Jerusalem? Measure this question in view of the numerous commandments, rituals, sacrifices that the Jews have made out of the temple service, or in fact the tabernacle service in the desert, laid down in the Torah. The Temple still existed at the close of biblical times, so do not expect an answer from Holy Writ.

Prayers composed during the Middle Ages are, however, full of this theme. The Temple is expected to rise again, and all its rites will be reinstated on the day of redemption. It is a nostalgic dream which I respect, though I must raise my own personal objections.

Exodus 25:8 says: "let them make me a sanctuary that I may dwell among them." This tabernacle has been built out by King Solomon many centuries later, the First Temple in Jerusalem as described in Kings 6.

The Second Temple was destroyed by the Romans and only the facing stones of its fundamental structure, the famous Wailing Wall, remained among them. All the concrete ritual acts and the tangible sacrifices had to be replaced by symbolic acts and prayers. We should ask ourselves if the outcome of the calamity was not a blessing in disguise.

For what should we pray? I do not ask to whom we are praying. We are facing His Holy Nothingness who created the world. That is why I believe that He should be symbolized as the emptiness of His essenceless existence, far away from the attributes of worldly ostentation. The tabernacle was carried through the dessert and did not need a proper roof. God ordered a cover of goat-hair curtains (Exodus 26:7). Why a roof at all? A cover serves the praying worshipers. The Holy Spirit does not need it. A *mikdash* is a sanctuary, not a comfortable gathering place. What we want is a dignified worshipping of as little symbol as possible. The rough stones where "Abraham built an alter there, and laid the wood in order to sacrifice his son," are quite enough. Even the pillar of stone on Beth El (Genesis 35:14) on this same place "where Jacob took the stones of that place and put them under his pillows" was too conspicuous. The rough stones under his head would have been quite enough. So why should the rough stones of the Wailing Wall not suit us—and Him—as a Holy Sanctuary that He may dwell among His people? In a temple we feast our eyes on aesthetics, but in prayer we should feast our mind on ethics—"the temple of the Lord, the temple of the Lord, the temple of the Lord are these" (Jeremiah 7:4). It is much harder for us mortals to build a sanctuary out of Nothing. Especially if we are obliged to write Nothing—nothing with a capital "N."

The Messiah has been a subject of study not among clear headed sages alone, but also among desultory obscurants. I am referring to the Kabalistic mystics. In the times when Christians waited in resignation and with folded hands for the end of the world. Some covens of the male sex were engaged in the study of the mysteries of holy conjuring books, such as the Zohar. They did this probably under the slogan, "if we are too stupid to find the truth through reason let us try madness." Most of it is entertaining, though not altogether innocent fiction.

These Kabalist sorcerers came to a conclusion that differed completely from the Christian view among the theologians. They told us that the world is split with God on the one side and Creation on the other. That our cosmos is an *olam ha'perud*—a split world waiting to be restored and to become *ha'olam ha'yichud*, the world unified by God. But human hands are required to complete God's work, a process which they call *tikun ha'olam*—universal restoration and unification. This view contains much which we intuitively recognize as an adequate conception. I ask myself, however, if these esoteric scarecrows did not try to burden us with an impossible task. How may mortals standing amidst the objects of His creation and being His creatures actively cooperate with God's Nothingness, and the hopelessness of this situation is not dissolved if we regard the created world as it is in reality, a mental world. God is absolutely transcendent, not only in relation to the world of matter and space but also in relation to the created world of mental reality.

When Lamentation Rabba 1:33 says that we are obliged to act as God's co-worker, we should read this with due modesty. The pretty fingers of human mortals cannot stretch into the great beyond to hold hands with the supposedly similar limbs of Holy Nothingness. What better may our Creator expect from us than that we promise to work within this world of Creation to improve it. Mystic guesswork does not help. Sitting cross-legged and staring at our naval will be of no use.

Notes

1. See N.N. Glatzer, *Modern Jewish Thought*, chapter on "Religious Postulates."
2. *Jews and Judaism in Crisis*.
3. He gave this in a short commentary on 20 February 1987 for Israel television.

16

Religious Eschatology, Part 2: The Christian Messiah

The fathers of Christianity were the fathers of the Christian Messiah, and the stupidities of the fathers are reflected in both. The circumstances of the first signs were a bad omen. Desperados, so-called "Essenes," fled by the hundreds into the desolate wilderness of the Judean desert. This was a long time before Jesus was born. They disavowed, as told, the service in the Temple because of the belief that the place had been irreparably defiled by Antiochus Epiphanes. The Holy Spirit had left the Temple and joined the devout refugees into barren desolation. They sang after they fled Jerusalem, "in the wilderness make a way for the Lord. Make straight in the Desert a highway for our God." They hoped that the anxious soul could finally relive itself in writing down scrolls full of new covenants with God.

They must have been oblivious of the plight of their mental state, which was the worst one could imagine for the promotion of healthy thought. What has been committed to the parchment scrolls, and what came to our knowledge, bears all the symptoms of troubled minds, of the neurosis from unbearable stresses.

The first sign of corruption of the sublime figure of the Messiah in Jewish folklore was the effect of the morbid feelings of nothingness of an inescapable fate that began to pervade some people under the rule of the Seleucides. In Greek this feeling is called *moira*, which means the absolute predestination of every step in one's life by the omnipotence of the Greek Gods on Mount Olympus. The Essenes were also mis-guided by a false interpretations of Proverbs 20:24, which reads: "Man's goings are of the Lord, how can a man then understand his own way?" and by Psalms 37:20: "The steps of a good man are ordered by the Lord and he delighteth in his way." These two quotations are adduced by Gaster's explanation of his translation of the Essene-scrolls in order to introduce us to the pessimistic spirit of these sects. But these two quotations cannot serve as valid arguments because their interpretation is wrong. The quotation from Proverbs means that the intellectual powers of God's creatures are limited, not that they are helpless. But the quotation for Psalm 37 simply means that God has shown us what is good and that He "delighteth in His way" after man has chosen all by himself between good and evil.

Let us have a close look at some sections from the Essene scrolls found in caves along the cliffs of the Dead Sea shore, known today as the Dead Sea Scrolls. They show how the meaning of the above Bible verses were distorted. "Before things came into existence He determined the plan of them; and when they fill their appointed roles, it is in accordance with this glorious design that they discharge their functions. Nothing can be changed. In His hand lies the government of all things."[1]

God provided man with two spirits, parallels to the later, Talmudic, *Yetser ha'tov* and the *Yetser ha'ra*, the "noble" and the "evil." But the Essenes believed that man was saddled with the evil consequences until the Day of Judgment and that he was not able to better his ways, nor to improve his conduct—during his life—and notwithstanding that the higher court of God would hold him responsible for all his misdeeds.

Here follows a section from the fifteenth hymn, 19–26: "Moreover, through the discernment which Thou has bestowed upon me I have come to know that not by the hand of flesh can a mortal order his way, neither can any man direct his own steps. I know that in Thy hand is the shaping of each man's spirit and here Thou didst create him, Thou didst ordain his works. And how can any man change what Thou has decreed?" This fatalistic doctrine emerged from the circumstances and the course of history in a time of upheaval that discouraged man to trust that he would be able to change the course of events. The Essenes tried to flee from this frustration, fleeing out of the city of Jerusalem, out of reality, but they could not stop reality from running along with them. Man becomes a robot in utter helplessness. He feels as if he had been conditioned to follow a foreordained path, even a sinful path, which must lead to the ultimate punishment on the Day of Judgment. He is the prototype of the mentally invalid on the conveyer belt.

We may pity the Essenes. We may blame the circumstances and specifically the evil Romans, but we cannot gloss over the evil consequences. The Essenes bore the responsibility of having created the frightening hell of a doomsday religion from destroyed Jewish faith—

Christianity. The man sitting on the conveyer belt, his face turned towards the disappearing past, his back towards the future, has not eyes in his occiput, but he is so preoccupied with the unknown that he lives under the suggestion that he is nevertheless able to see the occult. Nonexistent eyes peering into utter darkness of the unknown see nonexistent horrors. The vivid descriptions of the future in the Revelation of St. John the Divine, the last book of the New Testament, are a telling example. It introduces us in a very suggestive way into the unhealthy apocalyptic sphere of a pathologic fatalist. It is a work composed of Gnostic hallucination; it has been written by a poet, possibly under the influence of narcotics—which evoked visions of the forthcoming fall of the Roman Empire.

Because of this suggestive force it had become a central piece in the Christian religion, and especially among suggestible, simple people. The Revelation has therefore to be dealt with once more in the chapter on eschatology and apocalyptic revelation in psychological science.

Let us first admit that many great works of art before and after the writing of The Revelation have been conceived under the mood of utter distress. Its poet fled to a hermitage, allegedly to a cave on the treeless island of Patmos, in order to flee from his obsessive vision he tried to exorcise by writing everything down. He was a second Empedocles who had, five centuries before him, allegedly jumped into the crater of Mt. Etna in a fit of utter madness.

One cannot expect that the truth may ever flow from the phone in the hand of such a person and under such circumstances. I recall only one exception and that was Georg Wilhelm Hegel who managed to compose his clear-minded works under the smoke of gunpowder from the Battle of Jena in 1806.

As to the text of The Revelation itself: Obscure language in an epos that has been proclaimed Holy Writ engenders prejudices that soon turn into holy faith. Even some modern encyclopedias, such as *Meyers*, interpret the last book of the New Testament as forecasting the imminent fall of Jerusalem and the Temple of the wicked Jews. I call this a double lie, a malicious misinterpretation of a false prophecy. The style and view of narration has been borrowed from Daniel's dream (Daniel 7) which may allude to the despoiler of the Jewish religion (167 B.C.), Antiochus Epiphanes though the play on the stage refers to King Nebuchadnezzar in the year 586 B.C. and the destruction of the First Temple.

The Revelation of St. John refers clearly to the fall of Rome. Verse 17:3 says: "And I saw a woman sit upon a scarlet colored beast, full of

the names of blasphemy, having seven heads—," and a few verses further on 17:9: "The seven heads are seven mountains on which the woman
sitteth." Rome, not Jerusalem "sitteth" on seven mountains. Jerusalem
sits on only one mountain, Holy Moriah. Jerusalem is referred to as
being surrounded by mountains.

The clergy has always identified the Antichrist with the Jews, and
Thomas Aquinas was the ringleader. Thus it happened that The Revelation became the charter of the medieval anti-Semites.

Norman Cohen wrote a detailed compendium—*The Pursuit of the
Millennium*—of the emergence of Christian eschatology. "Millennium"
means the doctrine of the thousand years during which Christ is to reign
on earth according to Revelations 20: 1–7. The Greek word for this
doctrine is "*Chiliasm.*" The following picture emerged: the Jews who
live by human flesh and cadavers of the murdered, especially children,
will be led by the Antichrist, born in Babylon, back to the Holy Land
where they will be slaughtered en masse and burned. Though the Holy
Land did not figure anymore in the following centuries, the burnings of
the Jews became the inspiration for the inventors of Hitler's ovens. I
know from bitter experience that these pathological ideas predated Hitler
and his regime. The obsession to burn Jews alive was very much in
evidence among the Christian population in Holland, though it needed
a Hitler to carry it out.

I remember very clearly a most telling incident once among my cute
Christian playmates some time between 1920 and 1925. They whis-
pered in my ear that there are special planes on the aviation exposition
grounds in Rotterdam meant for the Jews. These machines will carry
them aloft and pour all of them strait into the hell-fire. Two among my
friends—gentile cousins—confirmed it, one got his story from his father, and the other mentioned a dominie at the Sunday school. I fear
that one of the sources is a verse in Revelations 20:15, "and whoever
was not found written in the book of life was cast into the lake of fire."
The European Jews caught in World War II were obviously not found
written in the book of life.

There lived in the early Middle Ages among the simple folk a quite
different type of Messiah, the opposite of the suffering evangelic one, a
redeemer emerging from the Sibyllic Oracles, an adaptation from an
old Jewish writing that had no other aim but to convert Greek pagans to
Judaism. The work had been renovated in the course of history into an
epic drama with a heroic warring Messiah wielding a flaming sword.
This was a figure apt to excite the imagination of the slaves of the

vagabonds thieves, highway robbers, and whores—simple folks do not like abstractions. They promptly identified the legendary hero with an actual mighty ruler. The Crusader, King Baldwin I (1058–1119), became one of the victims of popular imagination. He called himself King of Jerusalem, though he could rule his kingdom only from his coastal forts. After he fell in the dunes of northern Sinai, he rose soon from his grave in the imagination of his people, but the swindler who passed himself off as the royal redeemer was promptly unmasked.

One of the next kings to be elected by the fancy of his poorer subjects as the true Messiah was the Italian King of Germany, Fredrick II (1194–1250)—soldier, poet, legislator, and artist, the fanciful and fancy-inspiring. His majesty rose at least three times from this grave but all the impostors were caught. The people on the street joined zealously in the game and regarded itself part and parcel of the cast ready to bear a hand in God's work to exorcise the Devil, alas the Jews, and to topple Babylon. These imposters and firebrands emerged in the main from the scum of the people. They did not fight exclusively against their own social misfortune, but, above all, against the established religious convictions. These forerunners of the Reformation were the real enemies of the ruling clergy. The masses called them "*prophetae*"; and their early representatives were Jesus and St. Paul. Their inevitable fall did—in general—not remove them definitely from the state. In many cases they got the mercy to be slightly demoted to the rank of "precursor," the harbinger of the real Messiah. Jesus was spared this disgrace because of the power of St. Paul's imagination. He identified Jesus on the cross with his Orphic hero, Zagreus, in Greek mythology.

All these "*prophetae*" must have been very charismatic figures. Many kinds of hysterical sects amassed around them, dissenting cults that spun off from the chaotic mother church. There were so many that even the detailed description of Norman Cohen cannot glory in completeness. This is why I have to restrict myself to a very few representative examples and to a few remarks on the typical atmosphere of the movements.

The Reformation did not put a stop to the rise of these cults, which had in common that they were impatient preachers of the imminent millennium. They all bore a single grudge against the religious establishments, whether against the Catholics before the Reformation or against the mother church and the Confessional Protestants later. Some of the leaders were intelligent and noble figures, such as Johann Hus, or silly fundamentalists, such as Cornelius Jansen (1585–1638). Some among them preached doctrines that entailed very tragic consequences, notwithstanding the noble intentions. John Ball, who died in 1381, is often regarded as the father of Marxism—the Millennium would be built on social justice. Ball became the leader of the infamous Wat Tyler Revolt of peasant England, the foretaste of the Red Revolution set the kingdom on fire and was quelled with difficulty. Cornelius Jansen (1585–1638) on the other hand, could be regarded as a latter-day Essene, preaching the hopelessness of man's wickedness and predestined fate.

Jansen was a Catholic and even became a bishop in 1636. His book *Augustinus, Sive Doctrina Sancti Augustine*, etc. was written in a Calvinist vein and provoked the ire of the Jesuits. If man is incurably fallible, the Pope, really worshipped by the Jesuits, must be fallible too. The Jansenistic *"Utrechtsche Kerk,"* by many regarded as Catholic in essence, cannot live in peace with the Holy See.

Let us go back to the Middle Ages. A most picturesque figure was a certain Italian, Joachim of Fiore (1145–1202), who preached that the history of the world passes through three stages, the periods of the Father, of the Son, and of the Holy Ghost. The Father reigned from the moment of his great act creating the world, and His son succeeded Him and ruled through the Holy See; but His reign would come to an end in 1260, which would be a safe half-century after the foreseeable death of Joachim, thus safe-guarding him from being eventually indicted for a false prophecy. The year 1260 would be the first of the Millennium under the reign of the Holy Ghost.

This prophecy caught the church in a difficult position, especially sixty years later. Here follow some embarrassing alternatives:

1. St. Augustine announced that the Millennium began with the reign of the Church and this contradicted the preaching of Joachim.
2. If the prophecy of Joachim would have been correct, why did nothing special happen in the year 1260?

The most awkward aspect was that at least three infallible popes had been taken in by Joachim of Fiore. They have become the object of general derision.

Let me draw your attention to a particularly venomous prediction that typifies the spirit of the fresh German climate. *The Hundred Chapters* is a work composed by an author who lived somewhere near the Upper Rhine in the sixteenth century. It contains views that were borrowed by Hitler when he wrote *Mein Kampf*. Not Rome but Mainz was the eternal city; not the Jews but the Germans were the Chosen People; neither Latin nor Hebrew are holy languages, but German. The commandments were written by an impostor called Moses. But it would not be long before the decadent world ruled by the Jews would be crushed under the German heel and the millennium, "the thousand-year Reich of the future" would begin.

Various chiliadic cults and sects continued to emerge and to flourish after the Reformation. Velentin Weigel's movement, the followers of Jacob Bohm, the Rosicrucians in the seventeenth and eighteenth centuries, all lay claim to various forms of occult knowledge about the Day of Judgment. Johann Albrecht Bengel (1687–1752) tried to play it safe like Joachim of Fiore. He scheduled the coming of the Millennium on a day in the summer of the year 1836 in order that he would not be embarrassed during his lifetime, in case that nothing would happen during that special season.

They all have one trait in common. They misuse the Bible as coffee-grounds, or a crystal ball, or tarot cards to read the future. This species is by no means extinct. I bought some time ago a beautifully illustrated brochure written by Joseph H. Hunting called "Israel, a Modern Miracle." Hunting composed this during the euphoria of the Six Day War in 1967. All the great events during the twentieth century in the Holy Land are allegedly predicted in the Bible. He followed the common reasoning of many of his kind. If all the great events of the past have been predicted in the Bible, why should not all the other prophecies come true?

Note

1. *The Manual of Discipline* III 13:IV, 26.

17

Religious Eschatology, Part 3: Apocalyptic Revelation in Psychological Science

Seen in the context of psychology, the messianic ideal is an embarrassing problem. We think along a rut: good works, reward, gratitude and crime, punishment, redemption from remorse. Let us first have a look at the first threesome in connection with the coming of the Messiah. It has been said that true good works are performed with the object not to get a reward. Furthermore, that good works bring the Messiah near. If the Messiah would be a reward, it would follow that good works, are performed in order not to be rewarded with the coming of the Messiah, which is a self-contradicting absurdity. In other words, the coming of the Messiah, though related to the performance of good words cannot be a reward. It must be something not related to everyday human ethics, and as man cannot look beyond ethics, it must be something supernatural that a mortal cannot perceive.

We are taught: perform all the good works you can, but waiting for the Messiah is none of your business—just as thinking about your birth and your death is none of your business.

Let us now have a look at the second threesome which is still harder to understand when we examine it in context with the Messiah. My words are "redemption *from remorse*" not redemption "*from sin.*" The Messiah does not redeem the sinner, though Christians believe that he does. What I mean is expressed in Exodus 34:7. Moses asked to look at God face to face and the request was refused, as Exodus 33:20 says: "They cannot not see My face, for shall no man see Me and live." Only Nothing can perceive the Nothing. And in the special verse, 34:7, God is defined as "keeping mercy for thousands, forgiving iniquity and transgression and sin and that will by no means clear the guilty."

The words "the guilty" are an invention of the translator, an invented addition. In the Hebrew text we find the words "*ve-nake lo yenake poked*—whom does He by no means clean" A sinner who displays remorse in order to be redeemed does not repent. He wants his redemption as a reward for this remorse, just as in the same way, the performer of good works should not ask for a reward.

Bygones may be bygones but history cannot be undone; even sins cannot be forgotten. But even remorse must once come to an end. Let us call this first part of this subchapter "the ethical aspect of the coming of the Messiah." The expectation has, however, another psychological aspect which is by no means ethical and is not nice at all. Let us first have a last cursory look on the distant and not so distant past, on this primitive prescientific age, when people still drew all their wisdom from the Bible, and let us see how they digested—until recently—the sensational miracles whether witnessed or prophesied by the narrator. These miracles are described as supernatural phenomena but in connection with the eschatological Day of Judgment they are blown up to cosmic dimensions, commensurate to the uniqueness of the great event. They are meant to induce contagious panic. The reader, awestruck by the hair-raising phantasms, can hardly contain his curiosity.

I could not find any record of research into the mental reactions on biblical miracles by professional psychologists. It is even hard to determine the limit where these descriptions are only meant as a warning of what may happen if man does not mend his ways and where they just make you shudder. We may, however, be convinced that the prophets of the Old Testament had no intention to appeal to the lust of sensation of the common listener, their message did not convey a premonition of the inevitable. Look at the last words of the Old Testament, Malachi 4:4: "Remember ye the law of Moses my servant..." ending in 4:6 with "lest I come and smite the earth with a curse." The Revelation of St. John the Divine, however, had been conceived in an entirely different vein.

It is of interest to study from where the spectacle of the Day of Judgment has been borrowed. I believe that a great many of its attributes were adapted from the descriptions in Exodus 19:18: "And Mount Sinai was altogether in smoke, because the Lord descended upon it in fire, and the whole mount quaked greatly." The great signs on this occasion are not just meant to inspire panic, they conveyed a greater message than just fear. They ordered us to bow our heads in reverence to the supreme majesty of God's gift, our moral *a priori* that God bestows on the human soul.

This *a priori*, the knowledge of good and evil, was once awaked in the garden of Eden through God's design. As a small tribe of Israel multiplied to become a people, it became high time that these individual notions should be reduced to order and be formulated in a constitutional code even before the people would set foot on the Promised Land. This majestic moment was the crucial event of the whole Old Testament. An event so great that God Himself had to show His hand, visible through great signs enveloping the presence of God's invisible Nothingness. Defined in modern terms, God manifested Himself though the singularity that marked Creation and the singularity is invisible by definition.[1]

God the Invisible met the people halfway by descending on the mountain and He commanded Moses to climb up to Him. The links in the chain of events rise from one climax to the next. One should have expected that Moses, fearlessly challenging the smoke and the "fire which drape the majesty of His emptiness," would stare—on top of the mountain—straight into God's Nothingness. Nothing of the kind. The climax was a message consisting of a list of dry commandments how man is supposed to behave towards his fellow men. Man can never reach higher.

How could the narrator of this divine story pass from this highest apex to the unavoidable anticlimax without breaking the tension? Moses had to come down to earth again but his descent has been softened by a shock-absorber. And so we read in Exodus 34:29 "—Moses wist not that the skin of his face shone; and they were afraid to come nigh to him." How could an ordinary man not be awestruck at the sight of the only one on earth who has all the contents of our moral obligations spelled out in his mind?

Why did I hark back to the culminating point within biblical history while the object of this chapter is the miraculous end of history? Because many of the details described in the great spectacle on Mount Sinai have been borrowed by the New Testament to describe other great events, especially the Last Judgment. Even the shine on Moses' countenance has been used in Jesus' transfiguration mentioned in Matthew 17:2 and Mark 9:3. The luster in the Gospels is a bit second-hand though, a bit overdone, something like a leftover from Moses when he descended from Mount Sinai. Furthermore it appeared on Jesus' figure at an unexplained, even arbitrary occasion, serving no other purpose than to exalt the wearer of this radiation.

In short: the miraculous phenomena in Exodus 19 and 34 have been used as a prototype because of their awe-inspiring overbearance. Man seems to be in need to be overborne, and a less scrupulous preacher may be tempted to misuse this human foible to obtrude unconvincing doctrines upon his listeners. The longing for sensation is, indeed, a very normal trait, though of dubious moral value. It may easily grow out to become an unhealthy interest in apocalyptic, eschatological fantasies. This passion emerges certainly from our id when it is loaded with an unmanageable excess of a special kind of energy, especially among people with a very lively fantasy, children in the main.

They played superman with robots and swallow up the well-known pictures on the television screen and other products from the commercial market. This rather violent and destructive leaning may be countered by making good use of their healthy superego. The good guys are fighting against the odds of the bad guys and the good guys are supposed to win. As young minds like to be tickled by fear, misinterpreted as courage, they feel themselves as being participants in these wars and, overcoming their fear, they range themselves on the side of the good guys.

But not all the boys are endowed with an equal share of moral virtue. In some of them the passion for violent action is much stronger than the wish that justice should prevail. In these cases the passions of the id outweigh the forces of the superego. Their fantasy may be so strong that the pictures, evoked by all these suggestive narrations, may be so vivid that they lose their initial purpose to entertain. The emotions are pent up to the point where they have to relieve themselves through an outlet of violent and destructive action.

Though the performance on the screen came to an end, the play must go on, and it can only go on in the world of reality. The boys loathe the anticlimax that they have to return to this earth. This is a very dangerous stage and society has to safeguard itself against its evil consequence. Only psychological enlightenment may prevent irreparable damage. Good boys feel the limits of the allowable by intuition, but their characters vary so much that they shouldn't rely on a common and a moral sense; they should be guided.

How does the third domain of the soul react, which is, as we recollect, the ego? Its function is always much more impassioned and detached from the passion of our id and the emotional factors of our superego. The ego serves the superego in supplying the rational data that make a moral judgment possible. But the ego may also serve the id in this longing for sensation, by supplying the building-stones of rationale in a story full of passions. The result is science fiction, a rather modern symptom. Scientists with an urge to compose fiction-dramas make eager use of their knowledge to write stories that carry the readers away on the wings of their fancy.

But there are other expressions of the ego and the id joining hands. The computer for instance, or better yet, the various games devoured by bright and agile kids who are invited to run wild on the screen through the crossfire of robots moving along unpredictable paths. That is all much more exciting than using the computer to solve all the dry problems of homework. These occupations are rather harmless, though one may become an addict. They have at least the positive aspect that they train the alertness of the youngsters. Morally, they are neutral.

One of the factors contributing to this special interest in thrillers is, as I said, undiluted panic. An animal in fear runs away without questioning if the danger is real or imaginary. It follows a vital instinct and in many cases it saves its life. A human hero does not run away, but he is not fearless. After a soldier gets his medals for bravery, he is asked: "Were you afraid?" The usual answer in private is, "Hell, I wet my pants."

Real bravery is the insight that one has to fulfill a moral duty against the odds of mortal danger. Alas, many acts of bravery are only part and parcel of the lust-for-sensation syndrome, of having your lust-for-fear unhealthily tickled by an irresponsible stunt. Nobody is free from apprehension, though. "Only fools have no fear," said the hero in Heinrich von Kleist's *Der Prinz von Homburg*.

The passion for sensation consists thus of at least two elements: passion for the exciting—instilled, for example, by the gigantic size of the object; and a passion for fear, instilled, for example, by a propensity to brinkmanship. This second propensity often goes hand in hand with a propensity to show off.

When children are not fed by their parents with tales about giants and dragons, they serve themselves. I have a young grandson who, at the age of four or five, knew the Latin names of the cretaceous reptiles by heart—but only of the big, scary ones. They were all there in his nursery, small plastic toys from the boxes of his breakfast cereals. I took him once to the dinosaur hall in the Museum of Natural History in New York. He pinned himself with his back against the wall, his arms apart and his dark eyes awestruck, fixed at the huge skeletons. Then he looked at the mattocks and pickaxes used by the early paleontologists in Dinosaur Park about a century ago and at the other finer chisels that were used to dig the fossil bones out of the rocks. The little fellow had a great admiration for the courage of these people, hunting those huge animals with these primitive weapons. He asked me: "Didn't they have proper shotguns?"

Since the visit to the museum, he lost interest in the stark reality of these dead bones. But then he began to collect pictures of monsters. All his letters were illustrated with the rough copies of a new ogre. We have to realize that this is an example of normal reactions. It is not so long ago that everyone had to rely on the descriptions in the Bible to quench his thirst for thrilling stories. These stories were quite often pure inventions that emerged from false interpretation of obscure passages in the text where the original meaning of Hebrew terms fell into oblivion.

We may regard the Septuagint as the most reliable Greek translation from the Hebrew Bible. The biblical atmosphere managed to survive the great upheavals in the first few centuries. But soon it became apparent that the translation of Holy Writ into the modern languages was increasingly stymied by loss of memory. The original meaning of many words had been lost. What are the rows of precious stones that adorned the breastplates of the high priest in Exodus 28? We do not know the original meaning of the Hebrew words anymore in the terms of modern gemology. Could the indigenous rocks of the Sinai yield something more in those days than at present? The nearest place where real diamonds could have been found was in Africa, south of the equator. Why should a clear and transparent quartz-crystal not have served the purpose?

It became even a problem to translate the terms in the animate world. Bill Clarke wrote in the *Jerusalem Post* of 24 April 1981 that it is to be deplored that the less fortunate, who had to read the Bible in Hebrew, must live in a commonplace world. This person never encounters a dragon and the allurements of the unicorn do not entice him. Jews regard dragons, as they appear in the Christian translations of the Bible, as mere hallucinations.

The most baffling confusion resulted from the translation of the Hebrew "tan," plural "tanim," erroneously identified with "tanim," plural "teninim." A *tan* is a jackal and a *tanin* is a Nile crocodile. The brook that rises from the southern spur of Mount Carmel and flows towards the dunes of Caesaria still bear the name *Nachal Hateninim,* or "the Crocodile Brook." Crocodiles vanished from this area in the nineteenth century.

As to the other animal—the *tan*—Israelis who happen to live in the countryside still hear them quite often at night. But ignorant translators were oblivious of the slight difference in the Hebrew spelling; the two common animals were fused into one, and a four-legged monster was created. Adorned with scales it was called a "great dragon" (Ezekiel 29:1) and it lay in the rivers of Egypt. Professional Bible translators promptly associated it with the great monsters mentioned in Isaiah 27:1, which haunted the deserts and the ruins giving innocent people the jitters by their earsplitting howl. Look at Micah 1:8 for an example. The English translation goes that "I will make a wailing sound like the dragons." It should read "I will make a wailing sound like the jackals."

The crossing of a jackal with a crocodile is not the only monstrosity that ignorant Bible-translators have created. The King James Bible mentions the unicorn in Numbers 24:8 where Bileam compares Israel with the animal world. This "animal" appears again in Job 39:9–11 and in Psalms 22:21 and 92:10. The term "unicorn" is a translation from the Hebrew "re'em," but in the Spanish translation re'em has been translated as "buffalo," or the English buffalo. I do not believe the Spanish translation is correct, but it is at least down to earth. The Hebrew word for buffalo is "te'o," not "re'em." Re'em must be another strong animal. Every youngster in Israel who shows not more than a slight interest in nature would answer that "a re'em is a white antelope with thin horns that are so closely implanted on its forehead that, seen from a distance, it looks as if it has only one horn in the middle of its forehead. It runs with tremendous speed and outruns, in fact, every other quadruped in the desert."

This *Oryx leucoryx* was reintroduced in its original habitat through the services of the Israel Nature Preserve. That the re'em should be identified with this antelope is not better than a very attractive guess. I can only think of one perfect corroboration, the discovery of an antique mosaic-floor with the image of an oryx and the name "*re'em*" in Hebrew beneath the picture.

Medieval pictures of the legendary unicorn have some resemblance with the oryx, though it seems to balance its solitary horn, resembling a sugarloaf, a bit awkwardly in the middle of its nose, somewhat like a white goat on a masked ball. The *re'em* is extolled in the Bible for the fabulous strength of its horn, but medieval tales highlight its tremendous speed.

The Crusaders seemed to have encountered the *re'em* again. These heavily cuirassed horsemen could not spur their portly steeds to such speed. And soon the people were talking about a virgin, the only creature on earth able to catch the elusive unicorn, and even to tame it. Who else could this virgin be but immaculate Maria? In the oldest paintings she is pictured sitting and caressing the head of the domesticated unicorn, which tickles her under her chin with its sharp tubercle on the bridge of its nose.

And who else could this snow-white animal be but Jesus himself, wounded in the hunting? It became an inspiring subject for many medieval artists. The world-famous series of tapestries, with the embroidered representation of the great game-hunting for the unicorn—officially unveiled on 8 January 1499—dedicated to the marriage of Queen Anne of England (1476–1514) with Louis XII (1462–1515) are partly on display in the Musée de Cluny in the center of Paris (one has to get used to the dimness of the room to admire them), and the other half hangs in The Cloisters in uptown New York.

The final scene is the most interesting from an allegorical point of view. The hunter, probably the symbol of the angel Gabriel, succeeded in giving the animal a decisive wound whereupon the Virgin managed to capture it and to fence it in. It stands there pictured in its encircling pen composed of three round hoops (symbol of Trinity?) which are supported by twelve posts (twelve apostles?). The pen is allegedly the medieval symbol of the motherly womb of the immaculate Virgin from which the unicorn, alias Jesus, was born.

Let us realize that this complex of allegorical associations has been brought about by an obviously faulty translation from the Hebrew Bible. We are all human and we all tend to read a mysterious meaning in an obscure passage of Holy Writ and the mysterious becomes soon an established mystery. Jews are bound to fall in this same trap just as Christians are bound to be befuddled. These mishaps are rarer in Jewish hermeneutics than in Christian interpretations. Jews did not translate their Bible and this would preclude confusion of tongues, though not absolutely. A strange coincidence has corrupted the interpretations of Isaiah 34:14, which says, as translated from the Hebrew in the King James Bible that "the wild beasts of the desert shall also meet with the wild beasts of the satyr shall cry to his fellow; the screech-owl shall also rest there and find herself a place of rest."

The *satyr* is, by the way, a Greek mythological figure that Isaiah certainly had not in mind, though the screech-owl is a very reasonable translation of the Hebrew word *liliet*. I found a very odd footnote applying to "screech-owl" in the Bible that read "or nightmonster." This footnote is nothing but literally a Babelian confusion of tongues. The

Jews exiled to Babylon found, among the heathens and in Mesopotamian folklore, a certain *liliet,* a female Assyrian demon. The confusion became so deeply rooted in talmudic times that at the end of the Middle Ages, the days of "Rashi," the monster *liliet,* had long witch-type hair that hung down to her wings. She was the first wife of Adam, who made life a burden to him.

God's own court of justice had to dissolve the marriage whereupon the demon menaced God's creation with an early mortality of all the infants. Only special amulets saved mankind from total extinction. The source of the English word "lullaby," which means peaceful cradle song, comes from the Latin *Lula abi*, meaning *liliet* or "out of my sight." The poor rabbi who concocted this story must have had one hellish marriage.[2]

This last example may confirm that the passion to read mysteries in an obscure text is not just a Christian weakness. It is a universal human one, like that of my grandson who invented the war between the paleontologists and the dinosaurs. Christians have only become a bit more susceptible to this form of corruption because of the through-and-through mystical character of their religion. They always look for an intermediary to bridge the gap that separates God the Transcendent from His creatures. Only superhuman beings could serve as the supporting pillars of this bridge—dragons, unicorns, God's son. In their search they were inspired by obscure words in the original Bible that were promptly associated with the miracles that must happen on the Last Day. This is the inevitable consequence when a Christian reads the Old Testament with the eyes and the mind of a believer in the New Testament and especially in its apocalyptic summary, the Revelation of St. John the Divine.

Let us have a closer look at the text. It is meant as "The Revelation of Jesus Christ which God gave unto him to show unto his servants this which must shortly come to pass" (Rev. 1:1) "for the time is at hand" (Rev. 1:3, 22:10). This book is not meant to sound the alarm—that all these horrors may actually happen if man does not mend his ways. No, it is too late for that. All these horrors are predestined fate, and the detailed description does not serve a better purpose but to instill mortal fears. The poet had only in mind to convey that Rome was rotten down to the very core. It was beyond the faculties of man and he lacked the resources to redeem the world with his own hands. Only God is able to put His house in order. Sinful man stood in the way of His great broom.

Chapters 9–20 are crammed with terrifying horrors as no other book in the Bible. All this befalls to man who does not play a role. It is a question of taste if we should admire this hellish spectacle, even if we detest it. We are not allowed to belittle the rights of an artist to write down whatever may cross his mind. We are not even sure who the real author of the Revelation was; we should leave it to the judgment of the specialists.

But we have the right to ask about the identity of the criminal idiot who decided to include the Revelation in the Bible. He must have been either deaf to the most simple rules of pedagogics or he canonized its text with a criminal intent.

On reviewing the psychological aspects of messianism, the first question asking for a decisive answer is if it is harmful to reach beyond the acme of human faculties. We all feel that we cannot reach the top of moral virtue. If we would try nonetheless, what about it? The answer is that even if we would try to jump above our head, in order to reach for something of a higher value than moral virtue, for an unknown realm, we may be dead sure that we lose our sense of orientation, that we cannot tell anymore the above from below and that, on the contrary, we may be sure that we do not rise but sink down into the realm of mysticism, of self-mortification, of sin.

We just found out that if we have to believe in the Messiah, his coming would not be a reward for our good works, but an event of supernatural value. The supernatural is a closed book. In the Old Testament the Messiah does not even bear this name or any other name. And this may serve to answer our question. Do not try to reach above the acme of human faculties. And this answer settles the ethical aspect of the Messiah, which lies in the realm of our superego.

The other psychological aspect of Judgment Day is the aspect of our passion for sensation and excitement. We may find a reference to this point in Genesis 19:17 "—look not behind thee," when, "the Lord rained upon Sodom and upon Gomorrah brimstone and fire."

Whoever feasts his eyes on such a spectacle to satisfy his lust for sensation becomes a pillar of salt. One is allowed to believe in the prospect of the Messiah, to associate his coming with the performance of good works though not as a reward for our good works. It is forbidden to tickle our sinful imagination, it is forbidden to be involved in infernal spectacles, lest they may inspire us to "smite the earth with a curse" by our own hands, whereas we should "turn the heart of the fathers to the children, and the heart of the children to their fathers" (Malachi 4:6).

Notes

1. See Paul Davies, "The Creation of the Universe"—*The Edge of Infinity*.
2. See Nathan Ausubel, *A Treasury of Jewish Folklore*.

18

Religious Eschatology,
Part 4: The Fate of the World in the...

Many scientists put the Bible on one side but not their passions for great sceneries. A library full of books has been written about the grandiose phenomena which are supposed to mark the big bang. These deductions are of vital interest in so far as the calculations are based on sound reasoning.

There are plenty of reliable data available to trace the history of the universe back to its most distant past. Forecasting is a much more difficult problem. The future of the world could only be predicted if we would have an infallible and complete information about the present state of the universe. In foretelling the future of the cosmos, we are not assisted by telescopes which can look far into the distance. Nor can they see the remote past, the very date of the universe's birth, the big bang.

It is a trait that cosmology has in common with history. It holds only some chronicles of the past but none of the future. That is why drawing a prognosis is the most difficult state in scientific research. Information of the present state must be complete, which means in practice that there is never enough information. Even if we would assume that once, in the future, we would have gathered all the knowledge that our fallible senses could ever provide, this kind of knowledge would only serve to build up the gestalt of the physical aspect of the world, which is not the world of mental reality.

We have seen that from even the most complete information from our senses much of the mental reality must be withheld. We may reasonably assume that our senses are only sensitive to the special kind of signals from the external world (which we deduced to be the world of mental reality) and this means that we must remain oblivious of all the other phenomena in this world of reality. We may compare this limitation of our sensor-faculties with the feelers of a radio-set that are unable to transmit very low and very high frequencies.

Furthermore, we should never underestimate the limitation of our mental faculties, such as our innate inability to accept causes that engender more than one effect. Our mind rejects that it would be impossible to predict which effect would be realized, though this is exactly the case in the subatomic world.

Our information from the external world is thus not only incomplete, it is also indirect. Scientists very rarely realize this, they keep there eyes glued to the keyhole of their senses and jump to conclusion in their predictions. These predictions must even fall short of rough estimates. Small wonder that the prophecies of modern scientists contradict one another.

Everyone uses his own scales to weigh the future effects of the most simple factors. Optimists tend to underrate negative factors and the pessimists overrate them. Doomsday-prophets are popular, they satisfy the passion for the doomsday sceneries. Here follows a list of factors that allegedly spell our doom, it is just an arbitrary pick from whatever this sort of scientist tries to sell.

1. Life must come to an end because every species of the animal world had been doomed to become extinct, whether by lack of food or by their loss of immunity against diseases.

2. We are altering our ecological habitat and this must bring the human species to its doom. The process is precipitated by the creation of a surplus of industrial carbon-dioxide that must deteriorate our climate.

3. We cut down the natural forests that provide us our oxygen.

4. We destroy the topsoil by overgrazing and herewith the substratum of agriculture.

5. After every organic cycle a new belt of mountains is added to the continents. Continents grow and the climate must become more and more desert-like.

6. The cause of the organic cycles is the periodical gliding of the continental plates under oceanic plates. At the end, all the continents must disappear and the oceans will enwrap every acre of the Earth's surface.

7. The heat of the earth's core, the vital energy below the crust, is radiated away into outer space. The earth cools down and cannot sustain us anymore.

8. We live in a regime of high mountain ranges that act as holes in the glass roof of a hothouse. Heat flows away into outer space, conducted along the slopes of the mountains. The surface of the earth is doomed to cool and we will enter a new ice age. Homo sapiens will only survive if he is sapiens enough.

9. Another prognosis in which man's fate is the same as (8) is that there is a periodic change of the distance between the Earth and Sun resulting in the coming ice age.

10. Or, the entire solar system orbits periodically around a distant gravity-center and enters periodically a region full of heat-absorbing cosmic dust. This is a cause of periodic ice ages.

11. Or, the situation could be worse than above. The planets do not orbit along a fixed and rigid path and slowly spiral outwards, away from the sun. Stone-dead Mars was closer to the Sun and teemed with life. Earth will befall this same fate.

12. Or the sun is cooling down.

13. The sun's outer shell must behave like other stars and it will explode into supernova. We on earth will naturally be fried alive.

14. Our doom will be the result of expanding cosmos and the distance between the galaxies. Energy will, at the end, be evenly distributed (the energy entropy tends to a maximum). Life means concentrated energy but all the processes that may lead to new energy concentrations have become nonexistent.

15. Expansion of the universe must slow down because gravity and its contracting effect must prevail. At the end the cosmos must be crunched down again into nothingness.

16. A comet far away in outer space will draw near. It will enter our atmosphere and poison us—if it doesn't first manage to blow us to bits.

17. Mankind will arm itself until it annihilates itself.

I could go on until the cows come home; and the level is often one step above cheap science fiction. But this does not mean that we have to dismiss all the arguments out of hand—and I am not talking theoretical armchair philosophy. I am thinking of practical measures, how to curb the influences of points 2,3,4, and especially 17. Our ethical *a priori* obliges us to heed these warnings. We are obliged to support clear-headed nature conservation, and to pay as much attention as possible to the ethical aspects of conflicts.

As to the factors we are unable to change, the aversion of many to think about them follows from the universal instinct that nobody is eager to die, that—if we must die—at least our issue should reap the fruits of our life. That we do not stop even at the next generation, because we do not want the world to perish—not in the near and not in the most distant future. We instinctively turn away from these depressing thoughts and it may help you if I refer to our innate ignorance. The future of the universe must remain a closed book, forever, because of the limitations of our faculties. It seems that our ignorance about our individual future is less absolute.

19

God and Moral Virtue

The religious philosophy emerging from all the evidence that a transcendental God created everything mental, and therefore everything physical as well, may be a sound one from a scientific point of view. It is not in the least, however, an appetizing and encouraging philosophy from a human point of view. It ignores the human condition and cannot soothe mankind's pains. And the view that God created the world in a distant past, but seems to have forsaken us on that very moment, is called deism. It is the malady of modern science—not a philosophy, but a curse. Fom a human point of view it may even be worse than stark atheism. An atheist may embrace life without delusions—he has no choice. But a deist is tantalized by the semblance of an alternative: "God exists, but He is in heaven." We are not able to look beyond this testimony of despair with the aid of our common sense, our reason, our intellect.

But it is our good fortune that our intellect is only part of our soul. The soul contains other domains that may help us out, the domains of our emotions. It is of course from a scientific point of view, a perfectly sound practice to study emotions with the instrument of our reason. This science, which aims to reveal causal connections between emotional feelings, is called psychology.

But emotions may also be viewed from within, not only studied by reason from without. We may apply an emotional value scale to emotions. The top of the scale marks the saintly good and the bottom the devilish evil. These extremes do not tell the correct truth. Truth and value belong even to different categories.

Let us first analyze the scientific, psychological background of common sense and emotions. Psychological research led us to the triads of Sigmund Freud. The three domains together composing mind are the ego, the id, and the superego, which Freud called *"ich, es und ueberich."*

Our ego is the site of our common sense and intellect, our id is the reservoir of energy that feeds our soul, and the superego is the seat of our real moral sense. Taken together, our soul resembles a factory turning out thoughts and emotions. It is the powerhouse, the engine, and the storehouse of memories, all in one.

Psychologists study the id and the superego scientifically through the eyes of their third domain, their ego. This is their way of detaching themselves from the turmoils of their own turbulent soul, which may influence their clear judgment. To regard a thunderstorm from the outside as a meteorological phenomenon is much more comfortable than the frightful experience of a direct contact. Psychologists prefer observing the id of their patients with objective eyes, but shun an introspection of their own soul for good reasons. Whatever the study of an id may yield is rarely of much moral value.

But when we cross the threshold of the superego, we enter a royal domain where judgments are couched in the terms of good and evil. The ego is subservient to the superego; reason has to provide the superego with all the facts and the possible motives before it can pass its moral judgment. But the ego does not interfere in the judgment.

But what is the criterion of a moral judgment, a moral act? Much nonsense has been written about this problem. Socrates (c. 469–399 B.C.) confounded correct and incorrect reasoning with good and evil thoughts. The Spanish Rabbi Moses ben Maimon, known as Maimonides (1135–1204), knew the essence of the difference, but failed to stick to the point. The hedonists confounded the general and personal benefit with the moral good. Immanuel Kant believed that a "good man is a dutiful one," but he could not define duty.

Finally Gerard Heymans in 1934 hit upon the correct answer. He wrote that "a man who regards his own interests on a level with the interests of his fellow men, and acts accordingly, complies with the highest ethical norm."

Heymans called this "the principle of objective behavior" or simply, "the principle of objectivity." But whom should we call a good man? Heymans studied this question too and he found that good people are very reticent about their own virtues, though he found out that all of them happened to pay their taxes on time:

"But why?" Heymans asked. "Do you have the benefit of the country in mind?"
"Not particularly," was the reply.
"Do you want to impress your tax collector?"
"Not at all."

So why then are good people such conscientious taxpayers? Heymans seemed to have a hard time squeezing the answer out, but finally they disclosed the truth. They were all truthful to themselves. They carried the principle of objectivity to the extreme and they were unable to lie, even to themselves. To conclude that they regarded their own interest on the level of their own interests sounds absurd, but I mean that they participate in all their acts as their own judges. Not, however, as the consequence of a good education, but out of their own good heart; out of the nature of their character.

This conclusion gives the lie to the common misconception that our moral behavior should be the product of society, not more than a useful application of our instinct of common self-preservation. On the contrary, the worst feeling is the feeling of remorse and the best feeling is our heartfelt sympathy with those who perform a good deed. This feeling surpasses all the best wishes for the welfare of our society. To respect our moral sense is a commitment.

A certain professor of ethics peppered his lectures with insults and abuse. When one of his pupils objected he replied: "To require me to be ethical is tantamount to requiring the professor in mathematics to be a triangle."

Indeed a teacher is not expected to identify with his or her subject. But there is one exception—the teacher of ethics. This person is bound to observe the rules of ethics not in the capacity of a teacher but in the capacity of a human being.

A good society may be conducive to good moral behavior, but individual moral behavior is the only condition binding individuals into a good society. And this brings us to the subject of character.

Susan Stebbing's *Philosophy of the Physicians* contains a thought-provoking passage: "The mistaken view is that we have not the power of altering our character. The correct view is that we are able to alter our character if we wish."

Sounds impressive, but it is a view that I must reject outright. Character is the total of distinctive marks. Character is sometimes defined as moral strength and we talk about "a man or woman of character." But character is not only moral strength. It describes—alas too often—moral weakness. Let us therefore stick to our definition and regard character as the distinctive level of moral perfection.

This definition turned the wish to alter our character into a hollow phrase. Nobody is able to alter his or her own character or the character of anyone else. But let us first of all preclude a misunderstanding. The definition of the term "character" presupposes the existence of unalterable traits. This is implied in the meaning of a distinctive mark. Alterable marks are not distinctive. The distinctive marks of facial traits by which we recognize and identify our friends and family members are a case in point.

When I hear people heave a sigh of remorse it turns out that they do not want to change their characters, rather they are ashamed of their behavior. They are aware that they have acted below the level of their characters and the causes of their misbehavior may be many things. They may have been afflicted by bad impulsive habits; they may have been ill and not been able to posses themselves; they may have been ignorant of a moral aspect and their eyes may have been opened too late (which may imply that they had a bad education) and they may have been under the influence of a dominating tyrannical figure, who kept them from acting in accordance with their moral sense. These are factors that are extraneous to their characters but had a negative influence on their actions. It may always be explained to them why they have failed and how they may avoid similar failures in the future.

But let us now take culprits who are not amenable to moral arguments. They suffer from *moral aphasia*, sophisticated words meaning a bad character. The best education fails to enhance sensitiveness to moral aspects. It is the task—even the duty—of an educator to strive after the ideal that the actions of his pupils do not fall below the level of their characters. It is tedious toil to peel out their true nature from below the husk of bad circumstances.

The level of a few is, however, so low that an appeal to moral arguments is useless. In that case society demands a substitute for moral education, which is the stick and the carrot as defined by law. The criminal code is written for bad characters as a deterrent.

Physiognomy and character are, respectively, the physical and mental distinctive marks by which people are identifiable. Many of these marks are hereditary, and we refer to the genes in chromosomes as the carriers of these marks. Though it is clear that neither physical physiognomy nor the mental character may be defined as half of the sum of the parental traits. The progeny has generally lost some of the parental traits and some new and unexpected traits mark a new identity.[1]

The close connection between genes and characters has been studied by Hans J. Eysenck[2] who went deeper into the interpretation of a statistical research by Johannes Langer of Munich, who found by chance that brothers and sisters of a criminal showing signs of criminality is somewhat less than in the cases when a twin is the criminal. But the chance that two members of identical twins are both criminals if one of them is criminal is 100 percent greater.

This brings us to the end of the general philosophical, rational aspects of the first part. All the rest is based on common emotions. After a short survey of what we know about morals, we have still to study how we feel about them, how we value them.

Our intuition rates moral virtue as the culmination-point in the human soul, and we regard the human soul as the culmination of Creation. We cannot reach higher. But the altitude of the summit is not measurable in units of any parameter, either in the vocabulary of science or in any degree. A long and tedious research revealed that we are, by and large, and all over the world, in agreement about "good and evil."

Notwithstanding distortions in judgment, induced by all kinds of social circumstances, a general consensus prevails about the meaning of these two words. This convinces me that moral sense is something very real. I do not believe that mankind lives under the spell of mass hallucinations where moral issues are at stake. The certainty that God created the world has made me teach the conviction that moral virtue is a divine gift.

I would even dare to state that moral virtue is the only concrete sign linking us to our creator. I, however, agree that the linking thread appears to be very thin. It is, for example, beyond me why we are all endowed with a different dose of amenability to moral arguments. What is the meaning of the sharp contrast between the levels of our characters? Is a culprit the innocent victim of his character? It is a strange fact of life that we rarely ask about the source of a bad character. On the contrary, we justify the judge who metes out punishment in accordance with the law. It illustrates that our common emotions cannot be reconciled with our common sense. It is an unsolvable problem and we do not bring it one step nearer to a solution when we renounce good works as a means of attaining divine justification and replace good works with so-called faith. It does not change our judgment of a bad character. May the culprits be damned whether they acted out of bad character or whether they were deficient in their faith.

Worst of all it is a normal procedure in many religions to project this paradox in our human nature on the meting out of punishment by the Supreme Being, though He is the one who created us.

This deep misunderstanding is the consequence of regarding God as endowed with supreme moral virtue and ignoring that He created everything, including moral virtue, out of nothing. It is a misunderstanding that would turn our Creator into our Judge. It is an absurdity as old as religion itself. Here it is, spelled out in Susan Stebbing's work *Philosophy and the Physicists*:

a. God created me,
b. I am morally responsible for my deeds,
c. God punishes me for my transgressions.

Susan Stebbing concluded that it was not her task to solve theological problems. But it is mine, and that is why I propose a reasonable answer.

Let us take stock of our discoveries. We know that a transcendent God who bears no positive attributes has created the world out of nothing. He has created us. We are moreover convinced that nothing surpasses moral virtue, the commitment to regard our own interests on a level with the interests of our fellow men. The combination of this knowledge and this conviction leads to our common feeling that moral virtue is a divine gift.

But to regard our Creator as our Judge is only a figment of our imagination. The following alternative sounds better: in as much we are permitted to talk about God's will, we may feel that with His divine gift (moral virtue) He "wants" us to comply with our moral obligations.

This new insight frees us from the absurd view that we should worship God, our Creator, as "Your Worship." We are equipped to be our own judges and to judge among ourselves. We are equipped with the divine presence that enables us to accomplish this. What does the term "divine presence" mean for me? The French expression *mettre en présence* means to bring something within the presence of the receiver. But I want to give this definition a twist. Why should we ignore the donor? The donor is not always present when he donates, but the object of his donation is that the receiver should also be reminded of his presence. And from this may follow that God is in a certain way present in His divine gift, that moral virtue should remind us of God.

Illityd Trethowan[3] commented: "The absoluteness of moral obligation, as I see it, is so far from being self-explanatory that if it were not made intelligible by being found in a metaphysical—and, in fact, theistic context—I should be greatly tempted to hand it over to the anthropologists and the psychologists."[4] Here we went the other way. We started to explain it through psychology and ended by referring it to the theistic context. Brian Davies comments on Trethowan's comment: "If one already has reason for believing in God independently of moral considerations one might well argue that there is some additional reason for thinking of the moral law for reference to God."[5]

And this was indeed the direction or our reasoning.

Notes

1. I have to enter into more details of this subject when we deal with Darwin later.
2. *Crime and Personality.*
3. *The Basis of Belief.*
4. Brian Davies, *An Introduction to the Philosophy of Religion.*
5. See Davies's chapter "Morality and Religion."

20

The Modern World against God

Two movements are hard at work to undermine our world—materialism and mysticism. Materialism is theoretically wrong. It is also harmful from a practical point of view. It teaches that the world of space and matter is the real world and that mind is only an epiphenomenon, which means a phantom, an empty apparition. Moral virtue, which is part and parcel of the human mind, would therefore be a hallucination too. This is how materialists reason that moral sense is nothing but a product of society, an illusion without any intrinsic value. However, we concluded that individuals would never be ready to gather into a society if the members would not be endowed with moral sense right from the outset. In short: materialism is not only a fallacy, it is a moral disaster. Communism, based on materialism, proves this.

Mysticism is a much more sophisticated evil. It is a product of the absolute idealistic outlook at the opposite side of materialism. I recall that we deemed absolute idealism only useful in as much as it teaches that the world of space and matter is nothing but the mental world observed through our senses. But we deemed it wrong in its ultimate consequence. Absolute idealism cannot be reconciled with a world created out of nothing because this would contradict Spinoza's slogan, *Deus sive Natura*, that God and nature should be regarded as one and the same.

Well, one may counter that if the absolute idealists believe in a false theory it is their business. Their ideas may be useless, but the question arises: are they harmful? The answer is that they are not just a bit harmful, they play havoc with the slightly unhinged and even with normal people.

If the mental world is proclaimed "God" in compliance with Spinoza's slogan, our own minds would be sparks of God's consuming fire that craves to be reunited with Him in a feast of conflagration. But before one decides to commit one's soul to this flame one should weigh the consequences. Even such an idealistic diehard, who is so utterly convinced that his God is within his reach, cannot deny that he knows little if anything about his objectives. How will he react after he may find out that his aim is in verity, his unification with nothingness. I hope that it will sober him up as soon as it dawns on him that his contemplation and meditation were the means of premeditated suicide, of mental suicide, though suicide all the same.

This malady is called the mystic outlook because it falls just short of utter blindness. Mystics are the morbid outgrowth of every religion. The Jews have the Kabala; Christians call it "the penetration of the cloud of darkness," or by many other names; and Islam has its Sufism. These are all different terms for a common aim through similar techniques of meditation and contemplation. The misuse of drugs is rampant, for they are meant to stupefy the mind. The notorious mystic Aldous Huxley, for example, was a dangerous narcomaniac. He died with a shot of "acid" at his request.

The historical source of mysticism is the Far East. Hundreds of millions of people dispensed years ago with their common sense, and were afflicted by a chronic disease called irrationalism. Fresh brooks tumble down from the high Asian mountains to unite and form the sluggish Ganges. Occasionally the waters are dammed up and become huge stagnant pools, and the backwaters become infested with toxic germs. This happened also to the mental state of populations from China to India. The morbific germs are the carriers of fatalism and mental suicide. Travelers from the west were quite often infected and they carried the disease to Europe. One such was Roman Philosopher Plotinus (205–270) whose journeys to the Far East resulted in disaster.

The physical source of this morbidity is a negative mental condition. It always brings the urge to flee from the reality of life, which induces the frustrated to plunge into a world of mysterious otherness. This state of mind does not necessarily foster religious feelings, and it is not always an inclination to dabble in religious speculations, which leads into mystics. Mystics stand independent of religion. We cannot even say that mysticism is a special trend in the different religions. It is more a germ of infection, an infliction from the outside.

It is however true that some religions are more susceptible than others. The most transcendent religions, Judaism among them, are somewhat more immune. The mystic movement among the Jews, the Kabala, never managed to efface rational thinking, though Judaism had once in the beginning of the sixteenth century to purge itself from a mystic madness—the curse of the false messiah Shabbatai Zevi. This stampede into a world of fantasy was the fruit of the bitter plight of the Jews in Spain.

Christianity, on the other hand, is thoroughly imbued with so-called mystic Neoplatonic imminence. One of the principal aspects is the theological consequence of "the ministry of the Holy Spirit in the person and life of Christ." The practical consequence of this dogma is the induction of an illusion in the believer that it is within his or her reach to participate in this divine endowment. These people mourn for Adam who lost his share in the Holy Spirit because of his "fall."

A pious Christian seeks the restitution of this precious good in his soul through an act of God's grace. Huldreich Zwingli called this "transubstantiation" of man as the *"restitutio in intergrum,"* which is the restitution of mankind's primeval state—the state of the faithful believer before the "fall."

But Zwingli was not satisfied with the ignition of the Holy Ghost in the soul of mankind. He was after a mystic total unification with God's Holy Being. It is the craving of the sparks to merge with God's consuming fire. It must be clear to everyone who understood my explanation what God is, or better, what He is not, that this Christian aspiration is a sinful blasphemy. The abyss separating the Creator from His creatures cannot be bridged.

But Zwingli read the Bible only through his muffled eyes. To him Genesis 2:7 became *"Das Einhauchen des Gottesgeistes in den Menschen,"* which means "the inspiration of the Spirit of God in Man." This is a malevolent distortion of the text, "breathed in his nostrils the breath of life."

Zwingli's distortion of Holy Writ is even worse than accepting doubtful Holy Writ. Muslims cannot help that the same event is described in the Koran as Allah's gift of *"min ruhi,"* which is "of My Spirit," and that the Almighty ordered His angels to prostrate themselves before His creature, Adam. This episode puts man unwittingly on a level with God, and I believe that it has been a source of a false mystic image, an illusion in line with Sufism.

But I found the most malicious wickedness in Joel Goldschmith's *The Art of Meditation*—a confession in black and white that our moral sense is an impediment, a barrier, between ourselves and the higher state of union with God. Whither does Goldschmith entice his readers to follow him into Hell? Do we read in his statement Satan's complaint that he cannot deliver a good soul to Abaddon? The mysticism preached by Goldschmith is a Christian mysticism. He wants us to believe that

moral virtue is even an impediment, which has to be removed by meditation in our search for the grace of God. This is even worse than the
general Christian dogma, "the un-merited grace of God," which still
implies that good works are a merit.

But if moral virtue should be regarded as a barrier between ourselves and the higher state of union with God, moral virtue would be
despicable. I believe that this degeneration has again its source in the
Far East. An impressive number of oriental movements preach goodness in vague terms. One cannot expect clear-cut definitions from these
sources. I found in Allen W. Watts's *The Way of Zen* an attitude match-
ing the obnoxious ideas of Joel Goldschmith. It is a remark of the Zen-
mystic Yuan-wu who justified stealing and looting.

The means to attain the ends of the mystic—which is the torpor of
the self, of the ego as well as of the superego—are manifold, but they
cannot be selective in their effects. It is impossible to numb the ego and
leave the Superego intact. The means are too blunt and to coarse.

Aldous Huxley listed the following seven remedies well approved
of through the ages: hypnosis, isolation, to subject oneself to laboratory experiments, drugs and alcohol, self-inflicted insomnia, flagella-
tions, and long fasting. All of these are recommended for those who are
not stupid enough to manage without the use of artificial aids. Administered with proper diligence they induce what Huxley calls "divine
revelation." Many facts of life are hard to bear, but oblivion cannot
cure them. Sleep is the means to refresh the mind for the next day.

Aldous Huxley was one of the most fascinating and articulate mystics of the modern world, and within the context of God. Today he
deserves from us more than a cursory glance at his writings. Let us
therefore have a closer look at what was the pathologic source of his
mental disease. When we scan the records of his life, described in his
biography shortly after his death in 1963 (*Moksha*—"Mystical"), we
discover that Huxley's predilection for soporific drugs intended to invoke "divine revelations," emerged only during the last years of his
life. I would have been guilty of character assassination if I would take
leave of the readers with the impressions of the terminal stage of
Huxley's madness.

One of the most representative works he wrote when he enjoyed
more than a mere semblance of sanity was *The Perennial Philosophy*.
Although I agree with only a few passages, it reveals the sources that
exerted such a strong attraction on Huxley's mind, and the minds of all
the others who are susceptible to phantasms and illustrations.

The nidus of mysticism is located, like so many other poisonous ideas, in eastern Asia, not in a well-marked geographic point of this vast continent but in the characteristic way of life and outlook. This outlook made its first appearance in the writings of the "Upanishads," the Hindu scriptures of the eighth century B.C. Their fuzzy thoughts express how the thinkers of the Far East do not apply that fine and accurate apparatus of the human mind, which we call our working intellect, in the same way as Western scientists. The Western use of the intellect had never been studied adequately before René Descartes. His adage, *"Je pense donc je suis,"* I think therefore I exist, had already been challenged in his own days by Ralph Cudworth[1] who countered René Descartes's view that thinking is always an act of clear conscious cogitation by saying "a sleeping geometrician hath at that time all his geometrical theorems some way in him," and the geometrician does not cease to exist when he falls asleep. In short, Sigmund Freud neither invented nor discovered the realm of our subconsciousness, and unconsciousness; he only explored and surveyed it. Its existence was already known long before.

When Westerners talk about "thinking" we are so used to Descartes's view that we are hardly aware that we in fact are talking about an intense concentration of our conscious ego on a special subject. We intuitively see to it that no discursive thoughts on any other subject, nor any digressions, will distract our mind. Thoughts about alien subjects are submerged below the threshold of our consciousness. Edward de Bono[2] called the linear cogitation along the lines of logical sequences and the causal nexus vertical thinking.

But the numerous thinkers among the vast population east of Iran use an entirely different technique. The Tao-Zen, for example,[3] and other oriental movements dispense with the western habit and argue that the cosmos is not linear but three-dimensional. They liken the Western technique of thinking to the effort of surveying the contents of a dark room by scanning it with just one narrow and linear light-beam. They counter that this way is at best a time and energy-consuming effort, a waste. They therefore widen their field of vision, stage by stage, with a gradual broadening of their light-beam of attention at the expense of distinctiveness.

The picture becomes inevitably more and more diffuse until nothing can be perceived with what Westerners regard as "clarity." This attitude explains the long exercises of the Eastern monks in blurred thinking until they reach their object: Thoughtless nothingness, the *wu-nien* of the Zen.

And here we encountered one of the possible deeper motives of Huxley's inclination to plunge into the murky depth of mysticism. His near blindness, which developed when he studied at Eton, made him look for a world of the indistinct and the ill-definable, in which he could feel at home and far removed from the hostile world of normal eyesight. To be blind, deaf, and dumb beyond the reach of imaginative contrivances is the highest point of Zen. This is where we have true blindness, true deafness, and true dumbness. This picture of the landscape seen by the blind is just one of the numerous quotations of Huxley in his *The Perennial Philosophy*, which is mainly composed of citations—adorned with his personal comments—from Far Eastern literature. Quotations from the Christian mystics are in the minority.

But let us return to the oriental techniques of thinking. I have to admit that I occasionally make use of a similar technique, though not quite so consciously. When I feel that my protracted concentration on a problem did not lead to a solution, that my sharp focusing of my light-beam only succeeded in burning a tiny hole in it, I decide that I'll sleep on it. We call this giving yourself a rest when you suffer from mental fatigue. And lo, not in my slumber of the night, but the next morning usually between five and six, when I feel refreshed and fully awake, the solution announces itself.

What exactly happened during the night I cannot tell, but Edward de Bono advocated a more intentional and less intense technique he called "lateral thinking." Arthur Koestler[4] called it "thinking aside." We push the object of our problem, for a moment, out of the focus of our attention and open the way to fertilizing associations and inspiring illustrations.

The wandering of my mind during the night's rest serves a practical purpose; but in Eastern philosophy these techniques have become an end in themselves with outrageous consequences. In this aimless, wandering of the mind it is better to travel than to arrive. And to travel does not mean a purposeful motion towards a contemplated end, but the sleep-inducing rocking of the vehicle.

> *A cart went on a sandy road*
> *The moon was bright, the way was broad*
> *Its driver slept in peace...*

And this popular Dutch song goes on to tell that the driver never came home. What the driver was doing is called "meditation" and "contemplation." Many psychologists[5] maintain that there is a principle difference between meditation and contemplation. During meditation the active consciousness is not altogether shut off at the beginning of the exercise, but a certain subject is consciously chosen as a point of departure. Even during the next stages, this subject remains more or less under the attention of the meditating person.

Contemplation in its absoluteness is also a totally free nexus of associations as practiced by the analyzing psychologist. Contemplation was the technique applied by the "Quietests," the followers of Miguel de Molinos (1640–1697): "Thinking without words and seeing without knowledge, the absolute stillness." It reminds us of the drug addict slogan—"God is love."

Jacques Bénigue Bossuet, the Jesuit plotter, managed to have Quietism banned by the Pope in 1699 on the grounds that a passive floating in Quietist contemplation may eventually lead to spontaneous commendatory thoughts about the Devil, who manages to enter unobtrusively the mind of the mystic.

Aldous Huxley resented very much the oppression of contemplative mysticism that began to flourish at the end of the Middle Ages. He believed that "the only known method that really works is that of the contemplative." Let us realize that neither mystics nor poets can ever see the truth they pursue rapture. Let us awaken the charioteer on the sandy road and tell him that clear-mindedness is the only mental condition on which he may rely.

Huxley was educated a Christian and has always remained loyal to Christian symbolism, though the essence of Christianity went against his grain. This transpires from the passage in his book, *The Perennial Philosophy*. I followed his words with the greatest and sincerest admiration.

Huxley scorned one of Martin Luther's letters, the section "Esto Peccator," though it faithfully expresses the implicit belief of the Catholic and the explicit preaching of the orthodox Protestants: "Be a sinner and sin strongly; but yet more strongly believe and rejoice in Christ; who is the conqueror of sin, death, and the world. So long as we are as we are, there must be sinning; this life is not the dwelling place of righteousness."

This was the message in Luther's letter for Protestant Christianity. It became its fundamental doctrine. Luther's words are also the cry of a desperate who is preoccupied with imaginary original sin; who is drugged by this auto-suggestion for so long that he begins to occupy himself with real sin. Finally the dam of moral restraint brakes asunder and he is unable to stem the flood of his crimes. And then Luther has the cheek to entrust his redeemer, Jesus Christ, with the overtime work of cleansing the overload of his cart full of sins at the tollhouse of heaven.

Huxley fought courageously against this "king of religion." He warned against the misuse of the doctrine "justification by faith" as an excuse and even an invitation to sin. But his warning came a bit late—after World War II, which saw it rubbed into the blockhead of Christian society. Adolf Hitler and his Lutheran *volk* eagerly listened to the words that "this world is not the dwelling place of righteousness."

But Huxley was apparently not aware that he contributed a great deal to encourage the carrying out of mental exercises that must enhance what he even called "invitation to sin." We read this, for example, from his quotation from a medieval mystic work he admired very much, *The Cloud of the Unknowing*.[6] In spiritual exercises the words "God" and "sin" should be repeated continuously, thus, "—and mean by SIN a lump, thou knowest never what, none other thing than thyself—" and further on in this text "thou must feel in some part of this foul stinking lump in sin." The endless repetition of the "japam" of a magic word, a "mantram," is the common technique in the Far East that dulls the mind by auto-suggestion. This technique is by no means innocent, but it is outright sinful to use "sin" as a mantram. On the contrary, if one has to make use of a mantram—and I am not at all in favor of this—one should use "moral virtue" as the repeated slogan. It may induce bad people to good behavior, though they may quite often be disappointed when it turns out that they had been unable to live up to their ideals.

Huxley openly misuses the technique of auto-suggestion where he endeavors to blur the difference between personality, selfness, and selfishness. At the end they become even identical. Personality, that is, selfness (i.e., the self) should be mortified because of its sinfulness. In this point he sees eye to eye with the eighteenth-century author William Law who said: "Your own self is your own Cain that murders your own Abel. For every action and motion of self has the spirit of the antichrist and murders the divine life within you." It is, of course, true that selfishness is our enemy, but hating your own selfishness should never be mutilated to become self-hatred.

But Huxley was totally blinded—not only physically but morally, where self-hatred was concerned—and he quoted St. Francis of Assisi with an incomprehensible approval: "Self-knowledge, leading to self-hatred and humility, is the condition of the love and knowledge of God." St. Francis's "knowledge of God" and Huxley's "message" that "man's final end, the purpose of his existence, is to love, know and be united with the imminent and transcendent Godhead" are both nonsense. God is the nothingness, He cannot be known. May the mystics who persist in their search, desist from their very vain effort.

Huxley had not the faintest idea about the sense of ethics. He attacks the moralists, but it is obvious he did not know what it meant to be a moralist. They worship not God, but their own ethical ideals inasmuch as they treat virtue as an end in itself. This is the tragedy of the misunderstanding dating from the Middle Ages. The thinkers, Christians and Jews alike, preached the use of moral virtue as nothing but a means to know God. But after a very profound introspection we realized that we all bear moral virtue as the one and only sign of God by which we are created in His image—if we may forgive this clumsy expression in Genesis. Moral behavior is always an end in itself, otherwise the behavior would not be moral.

Personal ethical values can't be ethical, for ethical values are universal. And as to moral behavior, good works have a legal tender that is honored worldwide, because they are of a divine origin. They are God's gift in action.

Huxley's concept of God is not less faulty: "For Christian mystics the ineffable, attributeless Godhead is manifested in a Trinity of Persons, of whom it is possible to predicate such human attributes as goodness, wisdom, mercy, and love; but in a super-eminent degree." How the Attributeless may be predicated with the attribute of Trinity, with human attributes, is foreign to me. I guess that Huxley had entangled himself in the reading of Maimonides and other medieval Gnostics.

It is now fairly easy to take stock of the subject mysticism, at least provisionally. Every sensible thinker arrives on his way to reason at a junction. Is God only transcendent, or is He also imminent? Either the Creator maintains His creation or He does not. Science will never find evidence of a divine maintenance-person. The Creator created all the maintaining means, all the maintaining forces, all the maintaining laws of nature, which enable His creation to care perfectly well for itself. Science is not in need of an arbitrary interference from above. A perfect engineer built a perfect machine that does not need supervision.

We have seen that this deist view, which satisfies reason, does not satisfy everyone. Though the truth is absolutely convincing, it is emotionally unacceptable. Our emotions, not our reason, tell us that the deistic world of reason must be incomplete. This criticism is strengthened by another truth from the realization that our body is an aspect of the part of our mind, which is perceivable through our senses. It must follow that all that is perceivable through our senses is mental. If mind is real, emotions must be real too, which does of course not mean that every emotion reflects reality. It only means that emotions are real mental processes. From this follows that we are compelled to listen to our emotions, to scan them, to judge which emotion expresses the truth and which emotion is the consequence of inadequate thinking. We have only one instrument that helps us to discriminate between the true and the false of our intuition—reason. Intuition aided by reason is generally an infallible instrument.

When we are emotionally dissatisfied with the judgment of the deist, we ask for the cooperation of our reason and our intuition to look for the missing element in the deistic *Weltanschauung*—the element that makes life meaningful (again an emotional requirement). It is an element that gives us the feeling that our existence has a significance. Terms like "value," "meaning," and "purpose" only make sense and are only intelligible in the context of thinking human beings and human relationships. As a result we have to look for the missing element within ourselves, by thorough introspection.

Reason tells us that it is absolutely impossible that we would bear within ourselves something of the divine Nothingness. Man cannot have God in his soul—or even a spark of God. Our intuition must bow to reason. It is equally inconceivable, however, that terms like "value," "meaning," and "purpose-of-life" would bear no relation to the act of creation, that these words would be mere hallucinations of the human mind. It would be an emotionally unacceptable proposition; in other words, it would go against the grain of our intuition

And this compels us to take the plunge into the depth of our soul, on a voyage of exploration we call introspection. We have to climb, from the depths of our ids, slowly and carefully the stairs of our value scale. The only torch to light the realm of darkness is our intuition, and I told you that the highest point of the value scale is our superego and our moral sense. Our intuition tells us that moral virtue is a providential present, an indulgence, a divine tribute. It is a gift from God, but it is not in any way a part of Him. No human being can reach higher; our reason and our intuition forbid it.

In short, all the facts compel us to believe that moral virtue is the highest value of which we know, and it has been providentially created with the act of Creation. It is impossible to argue about this truth because it is the fruit of our emotional intuition and not of our reason. Nevertheless, it is the ultimate conclusion of a man of reason.

There are, however, numerous intelligent people, thinkers as well as others, who lost their way in the darkness of their introspection. They lost their way because the candle of their intuition blew out. They lost their foothold and fluttered like bats—misguided by the will-o'-the-wisp. These are the true mystics who mistake the will-o'-the-wisp for the Spirit of God. But the majority are a swarm of small bats trying to follow the tortuous and aimless course of leading wise bats, Aldous Huxley among them. They all stood, more than once, on the highest step of the value scale, on the acme of moral virtue, but they were not aware of it. They tried to reach higher still; they tried to use moral virtue as a means to reach higher to that phosphorescent wandering light. A light neither God nor the Spirit of God, but rather a faulty attributeless image predicated with human attributes.

The absurdity to ascribe human attributes to God, though admitting that attributes are nonexistent, served the purpose to satisfy the need for a personal contact with the Creator. This implies that the Creator has to be a person. The mystic, being only human and having no other attributes at his disposal but human attributes, has to look into the mirror in order to complete the image of his idol. He adorns it with all the arbitrary perfections he believes have to be super-eminent, for he wants his idol to be the supreme authority for which he yearns. All these perfections have to be blown up to divine proportions, to superlatives. Some of these attributes attain dimensions that compel the mystic inventor to call them by a different name. Huxley found a term in the Christian vocabulary, "charity."[7] It means compassion, loving kindness, goodness, grace, and so forth; all emphasized and expressed in acts performed without the expectation of a reward. When we scrutinize them it turns out that all these terms grouped together as charity comprise what is called moral virtue. To call its divine superlative "charity" ensues the estrangement of one's own human moral virtue, which the mystic believes to see—though in a grotesque enlargement—in the product of his fancy. He becomes overawed by the extreme contrast with his own humble virtues; but he is not just humble by nature, he has humiliated himself in front of a figure he created himself.

The harm he is inflicting on himself is enormous. He cringes and throws himself down, prostrated before an ideal he created by his own fantasy. He compares his own nothingness with the Person and then hates himself. He is ashamed of himself because he knows that he cannot live up to the divine standards and perfections of his imagination.

This is the source of the morbid wish for self-annihilation, mortification of the self. Protestant mysticism is even worse. A look in the mirror does not only reveal one's perfections, the mirror also reflects one's sinful traits. The mystic tends to magnify his own reflected fault, together with his virtues, just out of lack of self-criticism. It has been the fate of Lutheran mystics.

Martin Luther was a vicious sadist and he writes about his own God as follows: "He seems to delight in the torture of the wretched and to be more deserving of hate than of love." This is indeed a close-up of Luther's own wickedness. Aldous Huxley admits himself that Calvinism and Lutheran faith are degenerates of true religion, but that his own mysticism is only slightly better. Is this premeditated self-humiliation still religion or is it a pernicious auto-suggestion? What is this auto-suggestion else, but the subconscious evocation of an ideal as a standard of imitation with the dim awareness that this divine standard can never be reached? Pursued in earnest, it must lead to frustration, to an inferiority complex, and to a miserable life. These are the practical consequences.

But what are the practical alternatives? There are two. The first one is a remedy worse than the malady. It is using a being, one of our own living flesh and blood, as a living idol. Many who killed God are in need of His substitute, man. It leads straight away to the idolatrous worshipping of a prophet, a dictator, and this leads to dictatorship, to fascism, to Nazism.

The last and only true remedy follows after the luminous idea that we are obliged to use the high standard itself as standard. The highest level on the moral scale, that of the true virtuous, is not a "being" who has to be worshipped. The moral standard is, on the contrary, a divine gift. The obligation of moral behavior may be called a commandment of the Divine Being. If we would be allowed to talk about God's will, His will could never mean that He commands us to be God not even to be like God. *Immatatio Dei* does not exist. We are not commanded to be like Nothing, nor to imitate Nothing, by which we are only bound to become nothing.

Jews would point to Micah 6:8: "What does the Lord require of thee, but to do justly and to love mercy..." It is because "He has showed thee, O man, what is good—" and the response must be "to walk humbly with thy God"; not to be or to be like God. This response is moral behavior and that is why Jews worship the divine code as the highest standard.

Notes

1. See Arthur Koestler, *The Act of Creation*.
2. *The Use of Lateral Thinking*.
3. Allan Watts, *The Way of Zen*.
4. *The Act of Creation*.
5. William Ralph Inge, *Christian Mysticism Considered in Eight Lectures*.
6. Written in the fourteenth century. The author is unknown.
7. "Charitas" in Latin and "chesed" in Hebrew.

21

The Question of Life and Death: The Medical Approach

When we think deep enough about the limitation of our sensory faculties, they may seem to be a blessing in disguise, for the alternative may well mean that there would be little hope that our soul would survive. We know from experience that brains and nerves die with the body and if our senses and our intellect would be so ideal and perfect that they could reflect every detail in the mental reality as a "physical process" it would follow that the death of our body and our brains would be the physical reflection of the death of our mind. Theory and practice teach us that this pessimistic view has to be dismissed.

Indeed, our senses are highly selective and conditioned to pick only very special messages from the external world; and even if all our senses work in unison, it would be very probable that the resulting gestalt of the external world must remain very incomplete and defective; much of it must remain in the dark. This means that a dying body is not the sense perception of a dying mind. Gerard Heymans adduced a strong additional argument that there must be some form of parallelism between the observed physical law of conservation of energy and a similar, though unobserved, law of conservation of energy in the world of mental reality.

If energy cannot disappear in any process in the physical world, it must follow that energy cannot disappear in the mental world either. If this correlation is justified, we have to conclude that mental energy is indestructible. Should death be regarded as a disconnection of the function of the mind from (what we observe as) the function of the brain? It seems that this theoretical conclusion is on its way to being confirmed by practical observations in the world of medicine.

The article "Life after Death" by Mary Ann O'Roark in the *Reader's Digest* of September 1981, is a summary of medical research meant for the general public. The essence of this article is as follows: a mental sensation (which has been put into words) finds its reflection in the physical world as an "engram" in the brain. This engram is thus the physical expression of a mental memory. Physicians found that as long as we are alive, every mental memory remains physically engraved in the brain as such an engram. But this engram may of course deteriorate after months or years because of the natural slow loss of memory.

The question is what happens at the brink of death and beyond. Psychiatrists and others physicians have initiated a systematical notation of a large number of testimonies (in 1980 already more than 1000) from people who were pronounced dead—for example, on the operating table—and were successfully resuscitated, thus having passed through a miraculous recovery. In these testimonies the patients described their experiences, if they had any, during the critical interval of being not alive according to the clinical point of view.

Though the striking conformity of all these testimonies is a most remarkable fact, one should not jump to conclusions. We should first of all realize that the deterioration of the mental processes—which the physiologists would observe as the deterioration of the brain-process in dying persons may follow a rather universal pattern and that we might expect a universal similarity of hallucinations during the terminal stage, caused by cerebral anoxia. The well-known philosopher and theologian Hans Kung warned, for example, that this general similarity between all these testimonies may indeed be explained as pointing to a fixed physiological process and that, therefore, the contents of these momentary experiences do not need to point to any reality in the external world. Kung relied, in his provisional verdict for his book *Eternal Life?*, on an older work, R.A. Moody's *Life After Life*.

My principal objection against Moody's work is that it appeared somewhat prematurely. It was conceived when systematic research was barely off the ground. The number of cases was still modest and the conclusions should therefore be regarded as tentative. Another flaw in his work is that it includes interviews with people who suffered nothing worse than a few minutes of mortal fear in a near-accident. It is obvious that such experiences should be kept out of thanatology and that they have their place in normal psychology. They tell us nothing about death but very much about hallucinations, about impaired reason during moments of panic. But O'Roark's paper took account of much additional material that had not been available to Moody.

What do all the experiences, on which this paper has been based, have in common? All the patients felt they were floating through a tunnel towards a light they could not reach because the medical team managed to bring about a miraculous recovery—their death was prevented. Many recorded a first contact with friends and relatives who died before them, this was accompanied by a feeling of peace and bliss.

One of the common arguments of the skeptics has been discounted. Modern sophisticated instruments at St. Luke's Hospital showed that a sufficient quantity of oxygen continued to safeguard normal function of the brain even during the most critical moments.

"Clinical death" was established by the practical absence of electrical pulses in the brain-functions, in some cases even for several hours. And the patients nevertheless returned from their lethal state to tell the doctors what they felt. This discovery should bring many skeptics to second thoughts. But some may maintain that this evidence is not yet conclusive. Let us therefore come to the crucial point. Many patients who had not the faintest idea about sophisticated medical science were nonetheless able to describe in detail how the medical staff went about resuscitating them. Oddly enough, the patients did not observe all these activities from their recumbent position on the operating table but described how they floated around the room looking at their own operation from various angles. Allegedly even blind people were able to see everything during this brief period. After their recovery they were as stone blind as ever before.

These are unexpected miracles that I only dare to explain by way of the supernatural, and, please do not ask me further questions. Let us first of all agree that the topic is of such a vital interest that we have to encourage further investigation by unbiased scientists, under the condition, of course, that they strictly obey the rules of medical ethics. If these conditions are fulfilled, the public has no right to interfere, not even with the best of intents. This means that I do not change one iota my statement that the day of our death is none of our business.

Those who feel that they have learned something new about eschatology may ask how this new knowledge should affect their future behavior. I have told you that it should not affect their future behavior at all. Whatever I may have supplied in these last sections is some information to keep you posted. Let me only add that the patients about whom we were talking came to the common conclusion that life is precious or "it is wrong to violate the natural order of things by ending one's own life deliberately.[1]

Note

1. See, for an updated list of all the well explained cases, Colin Wilson, *Afterlife, an Investigation of the Evidence for Life after Death.*

22

Bridging the Chasm between God and Mankind in Judaism

This chapter deals with Jewish religion and how it is, in my opinion, supposed to deal with the problems of justice and rituals. I am not willing to accept a modern definition of religion that would justify to talk about "somebody who makes football his religion." I am old-fashioned on this point and unable to break the connection that links religion to God. There are, however, some groups who call themselves Jews, even to use the term "Judaism," but disregard the common definition of Judaism—"Religion of the Jews with belief in one God," as we find it described in the *Concise Oxford Dictionary*.

The vice chairman of the Jewish Association for Secular Humanistic Judaism, Zev Katz, wrote on 3 March 1987 in the *Jerusalem Post* in an article titled "Ignorant Prejudices" that "our effort to present Judaism without a supernatural being, without a transcendental, extra terrestrial, divine source is as revolutionary today as the idea of the One and invisible God was in the time of universal polytheism and idolatry." I fear that their effort is indeed "revolutionary" inasmuch as they try to overturn the facts that scientists—old and modern—have firmly established: that the transcendental, extraterrestrial, divine source exists and that this is not just a postulated fact but an established, scientific discovery.

I admitted, however, that we run into difficulties when we try to relate Him to our own, everyday, life and to our general human problems. I could eventually imagine that a sane and enlightened person would hesitate to accept all the ritual of an established religion merely on the ground of his knowledge that God exists. But Zev Katz, who is obviously a follower of Sherwin Wine (whom we met as the preacher of Humanistic Judaism) told us exactly the opposite in this aforementioned article. "We are 'Tora-bound' out of our free choice," he said. To be "Tora-bound" is a moral commitment to heed the moral obligations, the commandments of the Torah.

To be "Tora-bound" out of free choice may also be formulated slightly differently—that Mr. Zev Katz is ready to regard the Torah as binding. This expresses a much stronger tie than the statement that the Torah is a cultural asset of the Jewish people. That this difference is very relevant in our course may follow from a comparison. A Greek is to a certain degree bound to accept the works of Homer as his cultural inheritance, but this does not oblige him to believe in all the gods that dwelled on the Olympus. We may even admit that his free choice to be "Odyssey-bound" has a certain moral aspect. He wants to show his loyalty as a Greek, and his free choice to be loyal implies a voluntary moral obligation. But this implies not at all that he would oblige himself to believe in the gods mentioned in the *Odyssey*.

But compared with my imaginary Greek, Katz has put himself in an odd position. The *Odyssey* is a morally neutral epic, while the Torah is nothing but moral, and this would only mean that Katz would be ready to confess that he is doubly "Tora-bound" out of a moral loyalty with the Jewish people and its cultural assets—just as that Greek. And secondly out of a moral loyalty with its ethical contents.

And this may explain why I believe that Mr. Katz's credo that he is "Tora-bound" must imply much more than just culturally bound. The second difference between Katz and that Greek is, you surely guessed it, the belief in the Greek gods and the belief in the God of Israel—the unassailable fact that the Greek gods do not exist but that the Creator of the world exists. As to this last point Zev Katz is obviously unaware of the findings of modern philosophy and science.

I admit, however, that I am leaving the realm of logical reasoning and entering the realm of emotions when I proceed with my argumentation. I feel, but cannot adduce any scientific proof, that irrefragable certitude of God's existence, based on old theoretical and new practical scientific discoveries on the one hand and the intuitive, but not the less unshakable notion that there is no higher value than moral value on the other hand, are closely interconnected. And this feeling compelled me, as I have pointed out so often, to regard moral virtue as God's special gratification and tribute to the human soul.

Would we merge Katz's view with mine and integrate the two, then we would obtain indeed a Judaism defined as "the religion of the Jews with belief in one God based on Mosaic teachings." This, however, is a definition orthodox Jewry would find incomplete because of its basic principle in lieu the rabbinic teachings of the Talmud. I have to explain later that I cannot regard Mosaic teachings or the Torah on a level with rabbinic teachings, though I hold these learned discussions in respect.

When we talk about moral obligation in everyday life we always look at the needs of our fellow men. Are there any obligations to God or does the unbridgeable chasm that separates us from the absolute Transcendent exempt us from a useless effort? Though we have to concern ourselves in these last chapters with this fundamental problem, I admit that whatever I have to say is of an emotional nature and that some of my feelings do not tally with everything that orthodox or even some liberal Jews would have to say. Look one moment at the immensity of the problem. When an engineer plans a bridge across a wide valley that he cannot span by one arch, he first surveys the width and the depth of the depression. But do not demand from me to make a study in depth of the chasm that separates the Creator from His creatures.

I remind you again of my opening sentence that I could not find a shred of scientific evidence that God—in His essence-less Existence—has ever cared for the cosmos (has ever cared for us) since He brought the universe into being. You would be bound to ask: "How could we ever be committed to God if it would be true that He is not committed to our welfare?" And in practice: "Why should we serve God if He has never been at our service?"

These questions imply a seemingly self-evident relation-mutuality between God and man that obviously does not exist, though the Bible is built on the supposition of such a covenant between the parties. This covenant is, however, consistently being misread by all the priests, by all the parsons, and by all the rabbis who obviously believe in a mutual moral obligation between God and man.

It seems to be too hard for Brown, Jones, and Robinson to understand that the Creator out of Nothing has created all the mental essences, including moral obligations. Another question is that they tend to feel moral obligations as a necessary burden; they have to be convinced that far from being a burden, the moral obligations should be felt as an edifying, divine gift, though its Creator cannot have it and does not need it. We illustrated this with the tailor who creates his own suit without shedding his profession.

In the same way we may imagine God as wrapping Himself in all kinds of essences of supreme perfections—the perfection of supreme moral virtue, for example—expressed in His eternal loving care. I told you that our imagination misleads. We are committing the error of endowing the attributeless God with attributes. In simple language: God has no moral obligations, but He has seen to it that we are morally obliged to Him.

This moral obligation, the only bridge that leads to God, has a one-way traffic sign with a mark of exclamation. This mark is a divine request to show our gratitude to Him without expecting any reward for our expression of thanks. May this enlighten us that a benefactor has to expect only one reward—satisfaction with his benefaction, though this satisfaction is not a vital condition.

This is exactly the point where I differ with the orthodoxy. God has endowed His creation with everything it needs. The rest is up to us. We have no right to ask for special favors from heaven. I believe that no spectacle is more undignified than the cringing devout person who beseeches God for His administrations of Justice. God has endowed us with all the tools to judge among ourselves. Nobody else will judge us better if we fail in our judgment. We do not serve God by praying for forgiveness, but rather we debase God's image and ourselves—made in His image. We are saying: "My Lord I have sinned, here I am, the nothing created by You. I failed to judge myself. Please relieve me of this job of judging myself. I put myself in Your hands. You created me and You should judge me." I hope that you finally understand the paradoxical absurdity of this last sentence.

The only dignified service to God is showing gratitude. As it is said that the dead do not sing the Glory of God (Psalm 115:17), I would in the first place express my gratitude for having the opportunity to express it, for being alive. I am well aware that a few would not agree with me that this is a well-grounded reason for gratitude. I am thinking of those who still suffer from extremely abysmal experiences and memories. For them I have no better remedy than referring to the Book of Job.

Let me refer the question of how we should express our gratitude, which means how we should serve God in future. But I give you a foretaste by some preliminary remarks. Rabbi Aryeh Carmell wrote a critical article against Humanistic Judaism in the *Jerusalem Post* of 24 March 1987, and it contains a very apt quotation, Isaiah 40:6–8: "All flesh is grass, and all the goodness thereof is as the flower of the field. The grass withereth, the flower fadeth; because the spirit of the Lord— but the word of our God shall stand forever."

When we do not refer all the moral virtue possess, "the goodness of all flesh" to our Creator and do not regard it as His divine gift, all our best intentions are of no avail; they wither and fade. This is not exactly what Rabbi Carmell had in mind, but it comes very close to it. Let me add that moral virtue in action includes the judgment of the acts of others. There is only one day in a year when we judge only ourselves: the Day of Atonement. On this day we beat only our own chest, not the chest of others.

The only meaningful effort to make life bearable is trying to make it meaningful. Man gropes instinctively back to his mysterious fountain-head, and he tries frantically to maintain the link with his past in the hope to reveal his unknown destination, which he believes to be connected with his coming into being. I revealed the common morbid aberrations in the chapter "What We Are Thinking on Every Subsequent Day of the Rest of our Life." At the end we find that this problem is just as vexed because the efforts to bring God within our reach are in vain.

Whatever I may have accomplished is only a first step. I started with the intuitive acknowledgment that moral virtue has a divine source. Though this acknowledgment makes us aware of our moral obligations, it narrowed that chasm not more than just a little. Other means are attempted by all the religions, such as the observance of ceremonial commandments. They are all illusionary. But if illusions are necessary to make life bearable I am all for illusions. Making life bearable is, after all, a moral obligation. I mean, of course, those illusions that are invented devices to bring God nearer to man and man closer to God. Many of them, though not all, are harmful. The observance of Christian ceremonial obligations are—in the main—harmful. Christianity tried to build a steppingstone midway between God and man. Logically Jesus would have become a demigod, but demigods had been outmoded, and are too blatantly polytheistic. Hence they tried the middle-way, because Christ *qua homo* would be a steppingstone too close to the human shore, and Christ *qua Deus* too close to God to be of any use. He became a hybrid monster of Chalcedon, both man and god. But Christ fails above all on moral grounds as a redeemer of sins. He made man's will and conduct meaningless, and this made life meaningless and unbearable.

But the Jews have some holy men who did not figure as God and who were, on the contrary, thoroughly human. These were the rabbis, the *chachamenu zichronam le'bracha*—our sages may they rest in peace. They were not like the saints worshipped by Catholics, but their followers regarded them patently suitable to narrow the gap as agents between God and man.

I was reminded of Rabbi Solomon Schechter's conversation with a certain lady who remarked that Judaism is good enough for the daily wear and tear of life, but that it is so sane and plausible it never made a demand on faith, with the consequence that it never produced a saint.[1] I read in her words a covert reproach that Judaism does not make any special efforts to close the gap between God and man. I cannot help but feel a certain congeniality with the opinion of the lady in question, but I do not regard the obvious truth of her remark as something negative. On the contrary, I see in it a double asset. As to the first one, I point out once again that faith—belief without evidence—must yield to knowledge based on evidence. In our case, the knowledge based on evidence that our Creator exists. It follows thus that my own Judaism is based on a religious philosophy that hardly needs a demand on faith.

The other asset is that Jews are not requested to believe; instead they are requested to act, which is to keep the commandments. In plain words: do this and don't do that, but do not ask questions mortals cannot answer. This behavior is, in fact, faith in the bridge that spans the gap. Close your eyes and act as if the bridge exists; you will not come tumbling down.

But how did Solomon Schechter react to the contention of the lady in the story that Judaism could not produce a saint? He simply said, "enthusiasm and mysticism are the very soil on which saintliness thrives." If saints were defined as mystic cranks I might agree, but I believe that the soil on which saintliness thrives is more fertile than mysticism. What, for example, about the soil of goodness? It is all a question of how to define saints and saintliness, and if we want to obviate too personal a definition, we would be obliged to undertake thorough etymological research. This would take us too far from our subject; moreover we cannot expect that it would reveal anything that has not been known before.

Though many theologians understand that God cannot be described in terms of knowable attributes, they never shrink from describing him in terms of knowable perfections. They are obliged to follow the example of the Bible. God's typical attribute is His holiness. I shall devote a few words on the meaning of "holy" and of "holiness" in the very last chapter in connection with Abraham Joshua Heschel. Originally it meant "set apart," "special," "untouchable," and later "morally perfect." Sanctity and saintliness are not exactly synonymous with holiness. God is holy but not saintly, though God is holy and a saint is a holy man. God is never called a saint, for to be a saintly is intimately human. The difference is even felt in the romance languages where only the derivatives of the Latin *"sanctus"* are in use. Though *"les saints martyrs"* and *"la sainte Trinité"* refer to moral saints and the Holy Trinity. Even the French would shudder to call Jesus *"qua deus,"* "St. Jesus" and God, "St. Dieu." The German and the Dutch would regard "St. Gott" and "St. Cristus" as blasphemous. It follows that notwithstanding the frantic efforts by Christianity to narrow the gap between God and man (efforts which are expressed by the usage of the same perfections as man and God) the glaring contrast remains very much alive in the use of some among the attributes, such as the prefix "St."

I do not quite understand the words "to hallow God's name" (Dutch and German *heilegen*). Could it mean "to keep His holy name clean of blasphemous use," for God is holy and cannot be made holy (to hallow means to make holy). The harvest of my browsing in the fields of comparative linguistics is rather meager as yet. But now we hit upon another world related to the Latin *"sanctus,"* the English "to sanctify." We encounter this word for the first time in Genesis 2:2: *"Va-yevarech elohim et-yom hashevii u-kidesh oto*—and God blessed the seventh day and sanctified it." The Hebrew *"u-kidesh"* has been translated as "and sanctified." *"Le'kadesh"* is indeed "to consecrate," "to hallow," "to dedicate," but the English "to sanctify" is "to consecrate," "to devote," "to make holy." As to the English Bible translation, I have no objections; it is the meaning in the text of the Bible that gives me considerable headaches, especially because of the spurious rationale that follows "because that in it He had rested (Shabbat) from all His work which God created and made (*bara elohim le'asot*). Did God take a breather? We shall see further on that this was indeed the meaning and intent of the text. It is the only interpretation and it does not deserve belief. It is the consequence of the rejectable anthropomorphic image of God-a-person in the minds of the primitive.

The Bible does not make sense if we ignore the primitive cultural background at the time when the Bible was set down. Though the meaning of the Bible is eternal, its wording has become unacceptable since it has been put on record in the only terms that were intelligible in those times. Hence we have to reread the Bible and look at it through our modern eyes. As it stands, the word of the Bible could only mean that God celebrated His own accomplishment just for His own pleasure, and this is exactly the interpretation I would never accept, but let me try to make the words intelligible for us.

"Le-kadesh," to "sanctify," could mean also to dedicate to God or

another sacred use. As we have rejected the view that God dedicated the seventh day to Himself, He must have dedicated it to another sacred use. But which one? God could not have dedicated the seventh day to Adam and Eve who were not requested to observe the Sabbath, nor do we hear of such a commandment of their progeny. We are left in the dark, generation after generation, about God's mysterious intentions and about the act of sanctifying the seventh day.

Then all of a sudden the dawn rises in Exodus 20:10, "but the seventh day is the Sabbath of the Lord thy God," and further on "thou shalt not work." The day crowning the six days of creation was dedicated in advance by the Creator to a people born in the distant future. Since the observance of the weekly day of rest became an actual custom, it functions as a timekeeper for the Jewish people, reminding it week after week of the momentous instant of Creation. The rhythm of the ever-occurring Sabbath is the heartbeat of the Jews. But it turned out to be more. In Exodus it transcends the function of a ceremonial custom, for it says in Exodus 31:17 "—and on the seventh day He rested and was refreshed." The Hebrew word is here "*va'yinafesh*"—"being refreshed." God was refreshed, obviously exhausted after six days of hard labor. To stress the enormous importance of being delivered from the endless daily toil of slavery and to emphasize the exhilarating relief after the miracle of the deliverance, they projected the exuberance on the Supreme Being, Who is supposed to lead and to set the example. His people follows and imitates. The regular succession of repose to refresh oneself becomes a moral commandment to be observed by big and small, the Jews and the Gentiles in His service; the domesticated animals.

That is why the Sabbath is the most socialist among the commandments. The two prongs of this dichotic aim—the ceremonial custom on the one hand and its moral intent on the other—are expressed in the opening lines of the benediction sanctifying the Sabbath on Friday evening. Freely translated from the Hebrew liturgy: "Blessed be You my Lord, our God, King of the universe, Who bequeaths us (the commandment) to commemorate the act of Creation, to do this in memory of the Exodus from Egypt." It is a ceremonial custom to commemorate the act of Creation; to do this in memory of the Exodus from Egypt is a moral act, as follows from the special emphasis in Exodus on the moral aspect of the commandment.

If the Sabbath is meant to narrow the chasm separating God from man, does it succeed in accomplishing the task? Many an agnostic would challenge me and ask how the observances of the Sabbath could do any better than any among the ceremonial customs in other religions that aim at narrowing the gap between God and man. I would point out that the first requisites for that purpose must be that the custom implies a moral objective. I would also ask: is there any other ceremonial custom with a higher and more universal moral object than the Sabbath?

Am I holding you in suspense? Are you eager to hear from me how to proceed in the engineering of the bridge to God, which should span the yawning abyss? If so, you are only dimly aware of what we are engaged in; for I do not intend to proceed at all. I know that it is senseless; mind that we are still standing on the edge of the precipice, still on the solid ground of our familiar world. But our feet are trembling, for we are aware that it is not the crossing we dread, but what awaits us when we get to the other side. What we have to anticipate is that over there is the *Nihil* out of which the universe has been created. I point out once again that we cannot contact the Nothing if we would not turn ourselves into nothing, and that if we would succeed in turning ourselves into nothing we may not even get the chance to contact the Nothing. In other words, we would be engaged in mental suicide, in mysticism, on the verge of committing the mortal sin of self-immolation.

In short, we will never find out if our efforts to come closer to God by the observance of the Sabbath are aiming at an illusory target or not. We have not the means to explore it. We are only sure that the way to heaven is paved with good works and that the way paved with good intentions alone leads to another place.

If you would be disappointed because I cannot adduce more convincing evidence that the observance of the Sabbath serves a divine purpose, I would repeat that if you need an illusion to make life bearable will you please observe the Sabbath. Do all this and act and do not ask questions. Standing on the hither side there is no way to explore the yonder side. Do not heed the words of preachers who try to lead you on. Mind the bats who follow the will-o'-the-wisp.

Their misfortune is that they are misled by mystics who want themselves across instead of wanting their message, their prayers, across. They labor under a delusion that the wings of their prayers would transport them to their Lord. They do not understand that prayers should be like unmanned satellites that are launched to answer questions. The explorer on the ground should not even imagine himself inside that vehicle. He does not leave the launching pad. This image should illustrate what we should never expect of prayers. I remind you once more that if God created a spiritual world, it is absurd that He should be part of it. His transcendence is an awkward barrier that prayers have to overcome.

How do the customary prayers of the Sabbath relate to the Sabbath as an institution? Originally the Sabbath was celebrated as a day set apart by special sacrifices. The people were supposed to attend these ceremonies on this festive day of rest. The prescribed formalities of the Sabbath became more and more intricate and demanding after the inauguration of the Temple. Originally, the slaughtering of only two lambs was requested (Numbers 28:9); this number was trebled later (Ezekiel 46:4). It is obvious that the master of ceremonies could not contain his desire to heighten the contrast between the Sabbath and the workdays. This urge did not abate after the fall of the Temple. After the destruction of the Temple the system of public prayers was instituted to substitute the Temple service.[2]

Indeed, it may be an eye-opener to outsiders that we hear very little of a standard liturgy in the Old Testament. The present prayer-book, the Sidure, and the prayer book for festivals, the Machzor, are post-biblical creations, though they are largely composed of Bible texts. We find in them the same propensity towards enhancing the special sphere that benefits the Sabbath by their introduction of distinctive prayers and the inserting of extra lines in the daily prayers, all this with the object to accentuate the scansion of rhythm of Sabbath-days interfacing the monotonous flux of the workdays.

Whereas on the one hand the Sabbath was being made distinct for the workdays, the workdays on the other hand were being made distinct for the Sabbath. The fifth book of Moses, Deuteronomy, commands every adult male (6:8 and 11:18) to bind the words of "sh'ma" "for a sign upon thine hand and they shall be as frontlets between thine eyes." This is the commandment "le'haniach tefilin"—to lay the phylacteries every workday morning but not on Sabbath. Why? Do not look for learned "pirushim," halakic "explanations," for there is only one—it is just another way to make the Sabbath stand out as the special day of the week.

Here follows the only rationale of which I can think. A people that claims to bear the responsibilities of being the messenger of the divine moral code, the Torah, is in the need of marks of recognition, marks setting it apart among the other nations. The preeminent mark of distinction is the keeping of the Sabbath. But this mark made such a deep impression among the believers of other religions that the commandment has been borrowed from the Jews. The Christians have a Lord's day too, hence the distinctive mark began to lose its effect.

Pious Jews, however, do not see in the Sabbath just something that marks the Jews out, but they see in it a binding force that holds the Chosen People together, molding it and solidifying it into a solid pillar that supports the bridge over the abyss that separates man from God. Less pious Jews would be ready to admit that they see the pillar but looking upward they cannot see the bridge, still less how the bridge could ever carry their prayers across. The pious Jews would answer that nobody can ever see that bridge, you have to believe in it. And when you believe in it you trust it. Some among them have to be admonished that the bridge is not a device to carry them to God, but rather it is meant to convey their prayers.

Science and psychology succeeded in establishing the facts that God, our Creator, exists and honors the intuition that moral virtue is His special endowment. This is an enormous accomplishment, but we are totally at loss how to proceed. We have to leave science behind us; we are adrift and everyone is left to his own devices. I am adrift with all the others and the note I add must be a very personal one. I am talking to you about some "bridge" on which all the religions rely without questioning, a situation maintained for thousands of years. The bridge, the device linking man to God, served a vital purpose. Without it there would have been nothing but utter despair. In hindsight we must admit that the "bridge" safeguarded the survival of humanity. Psychologically it has been the possibility of something without its eventuality, the promise being better than the object itself.

But now, in our days and times of modern science, our eyes are opened and we fear that there is not even a bridge under construction. We are only aware of the two supporting bridgeheads, the pillar of Nothing at the one side in the Beyond and the pillar of our ethical *a priori* on our side. Hence, the bridge is supposed to span two worlds of different categories, the world of Nothing and the world of Creation. Why should we overtax our faculties? Shouldn't we give up the attempt? These are very relevant questions in connection with the theme of this chapter, dealing with our behavior in the service of God in the future.

The above questions cannot be answered with a simple yes or no. Whether I would recommend or discourage the attempts to strive after contact with God depends on the kind of person who confronts me. The observant nonorthodox prays out of habit; his or her conduct is perfunctory and we should either advise such people to get out of the old groove, or we should beat some honesty into the humdrum performance. It is either the one way or the other.

There are among them many fools who have their eyes glued on the divinity of the Torah and ignore the divinity of their own heart. They follow the Torah blindly, and the great majority do not even argue with their own ethical *a priori*. The few who do—there are well-known academics among them—believe, for example, that the tallying of commandments in the Decalogue with their own ethical principles is mere coincidence. If that would be true, the Bible would be so chock-full of such coincidences that we would make fools of ourselves if we would try to maintain that they are mere coincidences. Edward de Bono, author of *Practical Thinking*, said: "The most characteristic feature of stupidity is not inability to think or lack of knowledge, but the certainty with which ideas are held." They would not even dare to confront their healthy moral sense with God's commands. Their point of view is not Jewish but Christian—God's commands take precedence, one's own moral sense is nothing but this sinful world. They point at the words in Genesis 22, on Abraham who loved God more than his own son, and was ready to sacrifice him on God's command. They do not understand that the hand of the angel taught Abraham that he was on the verge of committing a hideous crime, that he did not hear the voice of God but of the devil.

My alternative is obvious: the Torah is not adapted to our moral sense, the Torah *is* our moral sense. I go even further: the Torah is not our moral sense elevated and exalted to the divine level, our moral sense itself is divine in principle.

Wherever my words are seemingly at variance with the words of the Torah, the correct explanations are too obvious. And I do not care if my interpretations are in agreement with the Talmud; for I would never believe that the sages—may they rest in peace—were wiser and better informed than we. I have the cheek to propose alternatives, especially where readings upset the harmony between our Creator and the moral sense He created.

Whether you address your Lord in the words of the standard liturgy, or whether you are inventive enough to compose your own, Psalm 4:4 teaches how to fuel the vehicle carrying your prayer: "Commune with your own heart upon your bed and be still."

Your own heart, your own moral sense, is the bridgehead on the hither side. You cannot commune with it without having complete confidence in its structure, that which supports the bridge that has to carry your prayers àcross. If God's commandments would be heavier than the crushing strength of the bridgehead, our moral strength, the whole bridge would be reduced to powder. The commandments imposed on us from above cannot be heavier than the commandments of moral conduct, and they cannot be of a different kind. They are even one and the same. If you maintain the orthodox view that God relies on you, He must suppose that you rely on your own moral strength that He introduced into you. This means that you are supposed to be able to judge His commandments on your own moral base.

Let me conclude this with a little reminder note. The ways how to serve God depend on what we believe. It is not at all indifferent in what we believe. I disagree with Werner Trudwin's silencer (in *Licht vom Licht*) that Christians should serve God as good Christians, Muslims as *fidèle* Muslims, and Jews as true Jews. Trudwin revealed himself as a superficial Christian and peacemaker. He said something like "as Christians we should learn to believe what others believe. But I believe that our own belief is still the best belief to believe in."

True difference should not be varnished over. A faith at loggerheads with the principle that only good works redeem has to be rejected. A faith based on tritheism has to be rejected. A faith that burdens a divine figure with our sins has to be rejected. I believe in human dignity, and a faith that debases God's creatures by despising them as nothings has to be rejected.

Notes

1. See the first lines of Schechter's *Saints and Saintliness in Studies on Judaism.*
2. Rabbi Leon I. Yagod, *Jewish Values,* "Tradition."

23

On the Future of Human and Divine Justice in Judaism

We judged justice on two levels, the divine and the secular. Recall that I consistently regarded *ius naturale* as justice instilled in the human soul by its Creator, with its religious aspect, and *ius voluntarium* as the day-to-day practice of justice after consulting the civil code, a man-made derivative of justice adapted after the ever-changing circumstances. I shall now show you that this distinction is blurred in Jewish life and I will illustrate this with two concrete examples.

Let me begin by slightly modifying my definition of moral behavior. Its modern equivalent is Heymans's principle of objectivity that I consistently described as conduct according to the principle that one should regard one's own interests at a level with the legitimate interests of our neighbors. My English version has a serious flaw—when no one was looking I added the word "legitimate." If I would define "legitimate" as "morally admissible," I would be in trouble. I would have introduced the term "morally" in my definition of morals and the definiendum should never appear in the definiens.[1]

Though my so-called definition would still appeal to our common sense, it would not deserve to be called a definition. Instead of "interested" I had better use the term "well-being." Hence, the new formula would be: "Moral conduct is according to the principle that one should regard one's own well-being on a level with the well-being of one's fellow men." The term "well-being" obviates objections against my propensity to use the word "legitimate" with its moral interpretation.

One of the central issues of moral behavior is the truth and telling the truth. Observant Jews agree. "Truth" in Hebrew, "*emeth*," has even been identified with God. *Elohim Emeth*—God the Lord is Truth—is an expression borrowed by Judaism from Kabalistic myth and mystics. But should the commandment to tell the truth always be observed under all possible circumstances?

Let us tackle two examples. The first one is patently Jewish, the other one is seemingly of a more secular and universal character, but nonetheless also applicable on Jewish jurisprudence.

I have the first example from a play and I knew the two characters quite well. I am not sure if the narrator did not mix them up with the real ones in some remote Polish village long ago.

One Friday night more than fifty years ago, the postmaster of the village of Zihron Ya'acov on Mount Carmel received a telegram from Jerusalem. It turned out to contain the message of the expected death of the mother of the rabbi of the aforementioned village. The postmaster deciphered the Morse code and wrote it down in Hebrew longhand on the then-official mandatory form of His Majesty's Service; he then sealed the envelope and locked it away in his drawer.

The Sabbath went by uneventfully, the rabbi performed his duties in the synagogue, Ohel Ya'acov, as usual and the postmaster invited the rabbi over after the prayer of the Havdala, marking the end of the weekly day of rest.

After breaking the ice the postmaster said: "I received a message for you just on Friday evening and that is why I could not deliver it." The rabbi paled and said: "You do not need to tell me what it is, I understand." And the week-long period of prescribed mourning began at that moment for the rabbi, his family, and the entire community.

I have told this incident over and over again to many liberal Jews and Christians and they reacted frequently by feelings of outrage that the postmaster misbehaved, that he should have told the rabbi—without delay—what happened and so forth. But the whole story cannot be understood without creeping under the skin of Jewish orthodoxy.

On the authority of those versed in Jewish jurisdiction, the postmaster behaved exactly as prescribed by halacha—Talmudic jurisprudence. It is strictly forbidden to desecrate the holiness of the Sabbath by divulging a message that is unwelcome to the community, unless silence would endanger the safety of its members. The postmaster, an observant Jew, was obliged to withhold the bad news until after the Sabbath.

Let us now take another example that seems profane at a first glance, but that is not profane at all.

Gerard Heymans's *Einführung in the die Ethik* contains the following passage: "A man who doubted if he should reveal to his gravely ill wife that the doctors could not promise her a long future—that the shock of telling her might be fatal—got the following answer from the well-known German thinker Johann Gottlieb Fichte: 'When your wife would die from the truth, you let her die.'"

Fichte's cynical answer is a mockery of ethics. He equated moral behavior with truthfulness at the very instance when the well-being of the woman would imply that one must be obliged to withhold the truth, not just for twenty-four hours as in the former example, but indefinitely. Heymans wrote within a different context: "The duty to tell the truth may thus occasionally collide with other duties, and this duty has, sometimes, to yield and to be overruled by the other duties." The duty to save a life (by concealing the truth) overrules the duty to tell the truth.

Are these arguments religious or secular? Let us examine the two cases separately. An observant Jew regards the Talmud to be as holy as Holy Writ, but I regard the Talmud as the surviving Jewish civil code of the Middle Ages. The counterargument that this civil code is based on the Old Testament is besides the point.

We may also regard the first example from a strictly human moral, and thus apparently secular, point of view. To maintain that the postmaster would have ignored the feelings and therefore the well-being of the community if he would have divulged the contents of the telegram on Friday night, might not be besides the point in the eyes of the liberal.

And here I add my personal conviction. You know that I am unable to divorce moral arguments from religious ones because of my tenet that our moral sense is a gift from heaven. The answer to the question of whether the first example is a religious or secular one is therefore ambiguous.

Let us now analyze the other example. We all agree that Fichte's answer was not just wrong, it was plainly wicked—the opposite of moral behavior. But even if you would ignore my view that moral virtue is directly related to God you could not maintain that the case has not its Jewish religious aspects.

Leviticus 18:5 says: "Ye shall therefore keep my statutes and my ordinances which if a man do he shall live by them." Talmud, Yoma 85b and Sanhedrin 74a comment that nobody should die as a result of blindly observing them, which means that in our example the dying women should not be "served" with the truth-and-nothing-but-the-truth that will lead to her early death.

In order not to confound the issue more than I already have, I propose to use—in the next few pages—the term "*ius voluntarium*" strictly as meaning the civil and the criminal codes as enacted by the modern legislator, but I have to return to the above subject in my appraisal of the orthodox standpoints in the State of Israel.

Divine justice is not justice administered by the Divine Being—this is impossible—but justice passed by humans. Kung relied in his provisional verdict, in his book *Eternal Life?* on an older work, R.A. Moody's *Life After Life*.

Note

1. See Peter A. Angeles's *Dictionary of Philosophy* on how to formulate definitions.

24

Worship and Service in Prospect

Religion preceded religious philosophy, emotion preceded reason; such is the course of history from the early primitive to the modern society, as well as from the childhood of the individual to his adulthood. This explains why those who have chosen religion as their profession and are sound-minded enough to think adequately, were in the main disposed to superimpose the rational thinking of their adulthood on the childish ideas about God as they are taught in Sunday school. The result is theology to be defined as apologetic explanation of faith—often against better knowledge. I moved consistently in the opposite direction: whatever came to my knowledge through reasoning, I accepted without further questions. Whenever rational thinking brought me to the point where my conclusions seemed to be ultimate and final, I tried to proceed against the odds after I consulted my emotions. But I have been very critical and censorious in my choice among the multitude of emotions that populate my mind. And my intuition, which, in itself, is an emotion (because it is an untaught insight) told me that I should rely exclusively on my moral sense.

This is why I believe that worship and service in religion should be based on reason and moral sense, and in my own context on the religious philosophy I have unfolded for you in the proceeding pages. Worship and service is religion in practice, and its kernel is the act of praying. I feel that I may add something more about the frame of mind during prayer that may transpire from the few scattered remarks in the text. The least you may expect of me is that I mark out a rough guideline, and even this modest aim turned out to be a very difficult task. I intend furthermore to turn my closing chapter into a recapitulative refresher containing many principal points made in the preceding chapters. All this will be applied to the subject of this chapter.

Being absorbed in prayer may evoke so many beneficial as well as harmful emotions that we are seriously challenged to examine to what extent one is permitted to let oneself go, to let oneself to be carried away. It may even be sinful to forget oneself in prayer. This indicates that I am less concerned about the influence of prayer on your health than about the grave danger that your self-effacement, whether intentional or incidental, may impair your faculty of moral judgment. I regard this question as a very serious one in view of the divine source of moral virtue, which is for me an axiomatic truth.

I propose to take the sermons of a famous authentic modern Jewish preacher as my whistling-boy, and to chart a safe route around the dangerous reefs—shaky principles of dubious credibility. My prototype is Rabbi Abraham Joshua Heschel, a man of old Hassidic stock who died in 1972. He was more a romantic than an Orthodox, and he did not mind to lecture at the Jewish Theological Seminary of the Conservatives, and even at the Hebrew Union College of the Reform, though the contents of his sermons were often in an orthodox strain. He was regarded by an overwhelming majority of American Jewry as the spokesman for Judaism, especially as a result of his discussion with Pope Paul VI in 1971. I do not doubt that his intentions were noble and sincere, notwithstanding his total disrespect for rational thinking. I propose to give him his due and use his best work as a guideline—*Quest for God: Studies in Prayer and Symbolism.*

The ebullience of this anthology of sermons may be the consequence of his descent; though Hasidism always endeavored to curb its own spiritualism (this in contrast with the other offspring from Kabalism, the Sabbatians). Many a Hasid regards himself in his bouts of delirious mysticism as being above sin and not obliged to take any moral standard into account. Heschel was in this respect certainly not a fanatic. He tried to match his own moral goodness with the supposed goodness of the Supreme Being, though he drew God as a forbidden, graven image. I am aware that his God does not figure as a Person. "We do not communicate with God," so he said and he specified: "prayer is not a real relationship between person and person." Nonetheless he endowed his deity with nonexisting attributes—His "desire," His "will," His "goodness," in short, qualities that Heschel (like so many other simple believers borrowed from his own idealized self).

Let me pick out three of his statements—let me call them *a, b,* and *c*—which betray his inadequate, slightly mystic and primitively anthropomorphic image of God. The first one, statement *a*, goes as follows: "Indeed there is something which is far greater than my desire to pray, namely God's desire that I pray."

In the second one, statement *b,* he said, "There are people who are hesitant to take seriously the possibility of our knowing what the will of God demands of us." But Heschel added that the words of Micah 6: that God requires us "to do justice and to love kindness" are binding even in the eyes of these doubters. Heschel commented: "If we are ready to believe that God requires of me to do justice, is it more difficult to believe that God requires of us to be holy?"

And in the last one, statement *c,* he said: "Without the holy the good turns chaotic, without the good, beauty becomes accidental."

Let us now clear our minds of hazy sentimentality and explore Heschel's three statements in a clear-minded mood. We begin with statement *a.* Here Heschel tried to convince us that God earnestly and vehemently wishes that I pray and that He expects that His will be done, though we know that He created will (because He created time and timeless willing is nonsense). "Thy will to create 'will' be done." Does this not summarize the stupidity? It is a sentence as absurd as, for example, "God consulted His watch, waiting for the exact moment to create time."

I believe that this clarifies that we have to reject statements that imply the will of God. "Will," to be defined as the faculty by which a person decides, or conceives himself, deciding, upon an action (at a certain moment) and initiating (at a certain moment).[1]

I am nevertheless ready to judge Heschel's statement *a* with the maximum of goodwill I am able to muster. I do not deny that the intense wish to pray from my side may invoke within me a certain suggestion that my wish factually stimulates a response by the Supreme Being. I am even ready to admit that my subconscious desire justifies, and even sanctifies, actual praying. From a purely scientific point of view, however, we got acquainted with an exclusively and absolutely transcendent and impersonal Creator out of Nothing. We have no scientific evidence whatsoever of an eventually emanating intercession by the Creator after the moments of Creation. We cannot adduce any better argument for God's permanent intercession after this moment than our intense desire for it. But yearning is, after all, nothing but an emotion and even the strongest emotional conviction is not necessarily valid evidence in favor of the existence of anything we wish to exist.

Nonrational cogitation is, as a rule, a will-o'-the wisp, but there is an important exception to the rule—our moral considerations that lead to good works. We feel that they are reliable, though nonrational, and this feeling is stronger than our strongest conviction of God's intercession.

I explained that this is an emotional axiom that cannot be explained by rational reasoning. But God's attributeless Nothingness forbids us from ascribing to Him the desire that we pray.

Let us now turn to statement *b,* which is rather complex. The verse Micah 6:8, "What does the Lord require of thee, but to do justly and to love mercy?" may be interpreted in the same strain as my answer to statement *a.* His requirement is His will and essence-less God cannot require, but He created in us moral will. We require of ourselves to do justly and to love mercy. Moral judgment is universally binding and it is understandable, even excusable, that devout believers surmise that it is God Himself requiring directly of them to do justly. If you are ready to reread Micah 6:8 in this strain, all the paradoxes will disappear.

Let us now associate the last words of Heschel's statement *c* and refer Heschel's demand to be holy to my judgment of Albert Schweitzer's personality. I admitted that Schweitzer was a holy man but that my judgment has nothing to do with the verse in Leviticus 11:41–44 concluding with "…and you shall be holy, for I am holy." I explained that the meaning of the words "holy" and "holiness" is today not the same anymore as it had been in the distant past, that holiness of a human being became more and more identified with supreme moral conduct. But this implies that we must read the verses in Leviticus as they were understood at the time when they were revealed: You, My people, should be something special, because I am something special and you should not defile yourself.

Heschel's question, "If we are ready to believe that God requires of me to do justice, is it more difficult to believe that God requires of us to be holy" suggests that he interprets the word "holy" as we interpret it today—to do justice as a saint. If Holiness is the superlative of goodness, Heschel's question may be paraphrased as, "If God requires of me good moral conduct, is it more difficult to believe that God requires of me to be a saint?" The answer would be that it is of course more difficult, just as it is more difficult to jump two meters than one.

I tried in vain to make sense of Heschel's statement *c.* He confused the different categories of values, a subject I have to deal with further on.

Heschel's writings create the impression that he was engaged in off-hand shooting-from-the-hip. He roughly estimated the effects of his words on his mediocre listeners but he did not calculate if his statements stood to reason. He even despised reason. "To the contemporary physicist the world of sense perceptions is of no relevance whatsoever. The familiar world is abandoned for abstracts, graphs, equations…

Science is purely operational, concerned merely with the manipulation of symbols."

Let us compare this gaffe with Albert Einstein's definition of physics: "Science is the attempt to make the chaotic diversity of our sense perceptions correspond to a logically uniform system of thought...." The sense experiences are the given subject matter.[2]

Who told the truth here, the rabbi or the professor? Was not Einstein the expert in the methods of the contemporary physicists, and was not Rabbi Heschel just a poor ignorant who poked his nose in a world that was to him "terra incognita"? And if it would turn out that Heschel is not an ignorant but that he knew what he was talking about, could one not indict him of slander?

The rabbi's behavior is a riddle to me; how could a man of this standing, who had been honored and adored, disfigure his reputation by uttering such a malicious remark? We may guess that he was under the spell of his own Hasidic mysticism, that many decent but simple people among his eager audience who were hanging on his lips, were scared away from science and reason because of the limitations of their intellect.

The more emotional types among them, those who hankered after a spiritual stronghold or a protective armor against the cold external world, were ready to forfeit reason altogether. This mutual resonance between preacher and audience did not guide, neither the shepherd nor his flock, to the safety of Truth. And this brings me back to the opening words of this chapter. Does everyone succeed in overcoming the primitive emotions of his childhood? Is everybody able to muster enough patience to follow a discourse in religious philosophy? It is often too much of an effort to sustain rational thinking; it is much easier to free oneself from such an intellectual constraint. The result is that man is easily overwhelmed by emotions of dubious value and his sick imagination gets, too often, the better of his common sense.

I told you that history turned the image of God into a terrible ogre provided with spurious attributes borrowed from the characteristics of human mortals. I explained that man painted his image of God on an imaginary canvas on which he enlarged himself, a self-portrait of infinite dimensions in comparison with the naught of his own belittled self. Let me try to express this absurd act in mathematical terms. The size of man's gigantic Idol attained its infinitive through dividing the enlargement of his own picture on the screen by the supposed zeroes of himself. This zero-ness is nothing but the result of his preexisting inferiority complex. It is obvious that the result of his stupid act is that he is unaware that the size of the projection, of his God, must be zero, such is the—in this case fateful—law of projection on a screen, the so-called conformity-transformation. I fear that this may be the cause of why mystics are sometimes in doubt whether God is a huge Person or a big Zero, which has nothing in common with His Zeroness as the reasoned consequence of the *Creatio ex Nihilo* but with the outcome of the projection of their own annihilated self, the inferiority of their inferiority complex.

Rudolf Otto in his work *The Idea of the Holy* called it "Creature Consciousness" defined as the "emotion of a creature, submerged and overwhelmed by its own nothingness in contrast to that which is supreme above all creatures." When a charismatic person fosters this feeling by sustained exercise, he may gather a large group of followers who are infected by a contagious disease called mysticism. Heschel was ensnared in the same mental state: "Feeling becomes prayer in the moment in which we forget ourselves and become aware of God."

All these mystics cringe in submission in front of the superlatives of their own attributes they call God, weeping in prayer, whining, blushing with shame. All human dignity goes by the board.

They have chosen the wrong viewpoint, craving for the nonrational numinous is a non-starter. One nonrational experience leads to another and so on, the catena of subsequent pictures and images is arbitrary. Everyone is engaged in his own personal fantasies, every single person creates his own God, Who has nothing in common with the God of his neighbor. Otto was immensely impressed by Isaiah chapter 6: "I saw the Lord sitting upon a throne, high and lifted up, and his train filled the temple. Above it stood the seraphims: each one had six wings, with twain he covered his face, and with twain he covered his feet and with twain he did fly." And Otto commented: "If man does not feel what the numinous is, when he reads the sixth chapter of Isaiah (check), then no preaching, singing, telling in Luther's phrase, can avail him."

Maybe I am in Otto's eyes an incurable stoic. The verses do not evoke within me any feeling of the numinous. On the contrary, I do not doubt that Isaiah gave an accurate description of his vision, but a vision is a personal experience, in this case a personal experience of God—It is a matter of course that I am frequently overawed by the mere idea that the world has been Created out of Nothing, but these emotions do not lead to self-depreciation. They do not affect my common and moral sense. I know that I am liable to excessive emotions but I always try to keep my mental balance. As to self-effacement of the mystic, I believe

that his act of self-destruction is even sinful. It implies that he depreci-
ates his own Creator Who created him, the negligible nullity who cringes
in front of the "God" of his own making and imagination. He may
avoid this blasphemy by trying to compose himself when a religious
experience—or even more worldly events that arouse strong emotions—
may make him weep. And I am not talking here about brushing away a
tear, but about copious tears. I do not trust their benefit. Many believe
in the cathartic (meaning "cleansing") effect of profuse weeping, but
my experience with mentally depressed friends, and especially with
my own moods, taught me to be on my guard. I found the same misgivings expressed in Arthur Koestler's research that appeared in his book
The Act of Creation, in chapter 12, "The Logic of the Moist Eye." If the
irresistible flood of tears should accomplish a "catharsis" it does not
result in the cleansing of mind but in the mind being cleansed off and
out.

I remember quite well that whenever I have been so totally upset
that I have become a disconsolate wretch, that I felt that something
vital had been broken within me. There might have been plenty of reasons for my state of mind, but it is not a reasonable state of mind. It is
natural, though, to surrender to grief, but I found it neither advisable
nor effective. The mind is blotted out and one may ignore the well-
being of our fellow men. It may lame our moral sense. Through the
moist and misty eye the "I" becomes invisible, as if it no longer exists.
Our soul is drained and turned into an empty shell, a hulk at the mercy
of the flood filling the vale of tears to the brim. Our lacrimal glands
secreted a mass of liquid that is whipped up into the stormy gales our
emotions. We lose our bearing and in utter despair we plead for God to
take the helm. But does He hear our passionate prayer?

What about the words of prayer? The Book of Psalms, in Hebrew,
"Tehillim," are bundled up in a small Jewish prayer book to be consulted in the hours of distress. For the really pious they work like tear
gas, stimulate the glands, and when their faces are already tear-strained
by sorrow, the psalms worsen their state. When this happens, lay your
prayer book aside a moment and if you are not being consoled by others, try to console yourself. The good, old adage should be kept in mind:
"Let me first try and perform my moral duty, I will cry later."

I know that this rule of wisdom is not put into practice among some
Jews and still less among many of the observant, who even cultivate
tears and turn this into a lifelong habit to repent daily for having been
born in sin as the son of fallen Adam. Jews who are under the influence

of Christian habits and try to copy them, should be ordered to turn their life of daily penitence into one of gratitude for being alive. They could say grace more joyfully after their meals. As to the inevitable sorrows, time usually takes care of them. Jews are ordered to rise after the prescribed seven days of mourning and to try to participate again in their daily occupations. Life must go on.

There is, however, one cause of grief that you should never allow to be dulled by the lapse of time—honest compunction for a real sin of omission and commission. Remorse is a warning sign that should be heeded by making good without delay. And if it may come to pass that you remember your guilt too late, and if you feel an urge to repent on the Day of Atonement, I advise you to take stock on that day in a mood of sobriety.

The following prayer is the greatest pitfall in Jewish liturgy:

"My Lord, before I was born I was nothing. And now, after I have been born, it is as if I have not been born at all. Alive I am ashes, so the more after I will be dead. See, here I am standing before You as a cup full of shame and disgrace. Please, my Lord, Lord of my fathers, may it be Thy will that I will sin no more."[3]

The beauty of these gripping words tempts our sobriety. They are so thought-provoking that only a thorough clarification in a sober mood may satisfy our longing for tranquility of mind.

Numerous questions arise. Why, for example, are we taken aback when we arrive at this highly emotive verse in the prayer book of the Day of Atonement? I answer this question with another one. Did you really feel like a "cup full of shame and disgrace" when you entered the house of worship? I fear that only a very few would answer yes. All the others are a mass of hypocrites, who feigned guilt feelings. Their mouths professed that they were standing before the Lord "as a cup full of shame and disgrace," but in their hearts they knew that they are much better and for good reason. We all must have gone through agonizing moments of sinning in the last year, but our sins never fill the cup to the brim. Reciting the words of self-depreciation in a mood that does not harmonize with your actual state of mind is playing an act before God. It is impious, insincere, and profane.

A great many among the perfunctory worshipers are suddenly impressed by the poetic beauty of the prayer. They are moved to tears. Praying, however, is not reciting a poem, but the fulfillment of a higher duty. Praying and reciting a poem cannot be judged with the same value-scale. The aesthetic value of a prayer is hardly an essential accessory.

Beauty animates the devout concentration but to be enraptured by beauty is not at all the same as to be in ecstasy in the fulfillment of a divine duty. If it would be the same, we could leave the rendering of the above verse to an actor, who, again, puts up an act, playing before God.

Finally we arrive at the few for whom the prayer is obviously meant; for those who are tormented by sincere guilt feelings after they committed real offenses. We may quote Bernard J. Bamberger, from his *The Search for Jewish Theology*, on "Religion in Action": "Sin is not so much falling short of perfection as falling short of one's own moral potential." Everyone is endowed with his own moral level, his "character." Sinners suffer guilt because of their awareness that they behaved below the standard of their character. They are ashamed of themselves and want to repent. (Bad characters, by the way, never feel compunction, and they do not behave below the standard of their low and mean characters.) It is hard to repent and not to humiliate yourself. You humiliate yourself when you recite the above highlighted prayer. To humiliate yourself is not the same as to be humble. Sinners who humiliate themselves before God depress themselves, before the religious service begins, into a state of exaggerated guilt. In extreme cases they may even invent additional offenses they never committed. Their humility is artificial, not less artificial than the humility of a pious Christian who is taught to believe in Original Sin.

What, indeed, is the effect of the verses? The lines do not simply say: "I was nothing, I am nothing, I will be nothing," they say something worse. *"Notsarti"* does not mean "I have been born," but "I have been created." "Before He created me I was nothing, after He created me I am nothing, and I will always be nothing." It is a cry of self-hatred and, in fact, hatred of all fellow-worshipers who utter the same prayer. It is the cry of the misanthrope who hates his neighbors and himself.

We may guess what the object of the prayer is and why it overshot the mark. It is the prayer of those who have drawn up their balance sheets of their behavior and found out that they are in the red. The next step should be a supreme effort to make good. Your words express that you are in the pits, and the first condition to make good is that you climb out. But this is impossible when you are saying, "After I have been created it is if I have not been created at all." The words of the prayer "misanthropize" and lame the will to perform moral duties; their effect is the opposite of their intent. Worst of all, the prayer implies that God created nothings; it belittles the acme of God's creation, Man, to nothingness. It is the most impious of blasphemies.

I believe that the words at the end top it all: "May it be that I'll sin no more." God created human goodwill, the will to do good and to oppose evil. Use it. Your will is the instrument in your own hands and do not ask for the intervention of "God's will," whatever you may understand by "God's will."

It is good to make up the balance of your transgressions on the eve of Kol Nidrei, but it is not less essential to take stock of the prayers that you are obliged to recite. We have seen that the lines we examined in detail do not serve any positive purpose and that reciting them is harmful—because they are the sigh of the poet and express this person's own intimate feelings. I beg you not to identify the "I" in the poem with yourself. A poem is not a prayer and the words in this special poem inflict damage because the poem has been used as a prayer. I propose to cut the lines from the prayer-book and to relocate them to a poetry book.

All these admonitions are an appeal to some ultra-Orthodox and mainly to those who are susceptible to extreme mystic hallucinations. They fell prey to the double-faced monster of mysticism and fundamentalism. The two trends are far from synonymous. The mystic is, as a rule, a passive addict to religious hallucinations, while a fundamentalist is an aggressive bigot. We usually find that one and the same person has been afflicted by both, mysticism and fundamentalism.

But I am writing this in the twentieth century and these two ailments were already on the wane a few centuries ago, to make place for another one, atheism, the revolt in the eighteenth and nineteenth century against the religions of the Middle Ages and of the Reformation. Its child is today's atheist, infected by the philosophies of mechanism and neo-Darwinism; an atavistic heritage form the times of Marx and Darwin. The modern discoveries by Albert Einstein, Max Planck, and Edwin Hubble on the one hand and old truths of the Middle Ages from Philoponus to Saadyah Gaon, rediscovered by Herbert Davidson, Austrin Wolfson and William Lane Craig on the other hand, brought many a nonbeliever inexorably back to the awareness that God exists.

The cured do not need our help, but what about the anti-intellectual, emotional stubborns? In their hearts they do not believe in their own atheistic philosophy and it is much harder to convince them. And then there are those who suffer from a callosity of human feelings. Heschel believed in jerking tears from these indifferent robots. These soldiers who are the most liable to commit war crimes under the spell of a very wicked ideology are, alas, not an exception. In one of Heschel's chapters, "Be Afraid to Pray" he remarked, "If they have no tears to shed, let them yearn for tears, let them try to discover their heart—" Well that's like trying to wrench juice from a dried-up lemon.

Between these two extremes, between the mystic religious and the atheists, lies the great majority of indifferents, and above all the wavering, the embarrassed by the antinomy of rational knowledge and nonrational sentiments, of how to compromise the First Cause with our moral sense in such a way that one may accept life as it is, including aging and dying. Let them realize that God did not create a moral universe, the world of relations between humans where the laws of moral conduct must be applied. We are all in agreement that the book of moral laws is the central jewel in the crown of God's creation, and this is exactly the reason why we do not accept aging and dying. Man regards his mortality as a great injustice, as an intrusion of the laws governing the amoral universe into moral rules of mankind.

Religion tries to transcend this glaring dissonance between our sad experience on earth and the supposed but nonexisting world of cosmic justice by conceiving that we enter the real world of justice in the afterworld where God is pictured as the divine Supreme Judge, passing judgment on the conduct of the souls according to the rules of moral sense of His mortals. The absurdity that our Creator would also function as our Judge is simply ignored by all the great religions. I recall that Albert Einstein once commented that "In giving out punishment and rewards He would to a certain extent be passing judgment on Himself."

Einstein further explained that the main source of "the present-day conflicts between the spheres of religion and of science lie in the concept of a personal God." Only a personal God may act and judge; hence, a personal God is an absurdity, our alternative was and is that God is the Nothingness that created the universe out of Nothing.

The next question is what happened to the Nothingness after the act of Creation? The answer is obviously "nothing." For we cannot reasonable conceive that the created world superseded part of God's Nothingness, though the Lurian Kabalists toyed with this idea—they called it the doctrine of Tsimtsum. Nothing of Nothingness "disappeared" with God's act of creating a something, which is the universe.

We should never entertain vain hopes that we would ever be able to lift the shroud of mystery that conceals the secrets of God's Nothingness. We cannot see beyond Creation into the cause of Creation. Religion is supposed to fill this great void in our knowledge, but the faithful can only pray to whatever the Nothing may contain, he should never try to pray to what the Nothing cannot contain. The nothing cannot contain anything that is not Nothing such as the essences that we borrow from our familiar, created world. It is not of concern to him to presuppose an essence of the "to Whom" his prayers may concern. Though modern religion is generally in agreement on this point, it continues very often to ignore its pitfalls. It reasons more or less as follows: "Though the existence of God is an established fact, He cannot be approached by reason. (He is a fact in the orthodox churches as an ossified doctrine and according to modern liberal theology as a corroboration by philosophy and science.)

But we do not accept an unapproachable God, and, thus, we have no choice but to make Him approachable. So let us muster all our nonrational sentiments and draw from them His image in such a way that it satisfies our sentiments. Koestler expressed this in his terse way. He said that "theologians start from the premise that the mind of God is beyond human understanding and then proceed to explain how the mind of God works." But all my reasoning does not keep the devout from his quest for God—he prays.

Despite several sections in the preceding chapters dealing with prayer we must look at some other problems. We will endeavor to answer the following questions:

a. What to pray?
b. To Whom do we pray?
c. Do we expect anything from prayer?

The question "What to pray?" has already been answered. Our prayers may be petitionary, intercessory, or an expression of thanksgiving, but whatever the aim of the prayer we are reminded that prayers with a moral object should always take priority. This teaches us in what direction to guide prayers. It is written in Deuteronomy 6:5 that "thou shalt love the Lord thy God with all thine heart, and will all thy soul and with all thy might." We identified "heart," "soul," and "might" with the triad of Freud's superego, id, and ego and in this sequence. Our heart, the superego, comes first; it is the only reliable fountainhead from which good prayers may emerge. We do not pray with our cold, calculating ego, not with the wild emotions of our id if we are sensible enough not to fall into the trap of mysticism.

The second question, "To Whom do we pray," is an item of lively discussion since the nineteenth century. The crisis began with the discovery by the idealistic schools that matter can never create mind, that the old materialistic outlook had to be rejected. It became clear that even the most intricate pattern of brain-substances and the most complicated processes in the brains are unable to produce mind. Though this observation is obviously correct, it is only part of the truth.

The conclusion that the universe (as we observe it through our senses) is a translation of a reality that is not observed matter in space but a spiritual reality—the idealistic point of view—is not the ultimate conclusion. We have seen that another conclusion—that the observable world of matter in space has been created out of Nothing—must imply that the real, spiritual reality must have come into being out of Nothing. It is this last conclusion that is being ignored by many modern idealists who try to convince us that we pray to a special element or factor that is part and parcel of the spiritual reality. In their eyes, God and the world are one and the same. They reason that this special element or factor, which is the human goodness we regard as the most noble part of Creation, is worthy to be identified with God. They believe, in other words, that we actually pray to the goodness within ourselves. (The word Creation does not figure among these idealists.) I regard this view as the nadir-point of narcissism.

Rabbi A.J. Heschel was right. He commented on a religious discussion published in 1911 that it is a definition of prayer that fits in pantheism. "If the deity is equated with the universe, and we ourselves are part of the universe of the deity, we are praying to ourselves." Because we are not part of the deity but part of the universe created by the deity, human goodness is a creation. "The jewel in the crown," I called it; and we do not express our gratitude to the jewel, but to its donor.

I do not know of any prayer that surpasses thanksgivings, but Aldous Huxley, the notorious mystic/drug addict, disagreed. He called prayers of thanksgiving "adoration" and he rated them one step below "contemplation."[4] "Contemplation" stands for "thinking of God." When I think of God I think of Nothing and when I keep my mind hard and long enough on Nothing I feel either nothing or I may feel myself sinking into a state of nothingness. This is indeed the aim of the mystic, who nurses his nothingness in front of the Almighty. Would I sink so low, I would be unable to fulfill my moral duties; and that is hardly a proper state to express gratitude to the divine Donor of my moral virtue.

Finally, are we justified to expect any positive results from prayer? A learned rabbi once said: "God may answer our prayers; He never answers our questions." I suggest an amendment: God gave us the tools to answer some questions all by ourselves, but they are not adequate enough to solve many others. The less we discuss unsolvable questions, the wiser we are.

I may fancy that I address my petition to a Being that is not strictly a person in the human sense and nevertheless entertain the hope that my desire will be met. But it is rather awkward to expect any response from a Being devoid of any knowable essence, a being that I experience as Nothing, albeit a Nothing that is the cause of everything. In short, there is no soul in the world who can answer the question whether our prayers may be answered or not.

Nevertheless, I feel that our faith is justified. We would never accept that our intuition and our moral integrity are divine gifts, nothing but a baseless mass-hallucination. If moral integrity is indeed divine, it must make sense and from this must follow that prayers with a moral object must make sense. Making sense is however not the same as serving a purpose, for we have seen that purposes are imperfect human ideas. This brings me to an intuitive kind of insight that—I hope—does not deceive me. Moral sense transcends in some way the narrow field of relations between man and his fellow men—and this in spite of the glaring truth that God did not create a moral universe. This is a fact that has been such a serious stumbling block for many thinkers to believe in God—Bertrand Russell among them.

This may have been an edifying conclusion of all I have to say but I want to discuss two additional items: Jewish education and the commandments in the context of biblical history. Let me first address the rabbis and their attitudes, their settled mode of thinking in their confrontation with their bored flocks.

First of all, what is the countenance of a modern rabbi? He is a normal human being not very different from the average member of his flock. He stands out, however, in his acquaintance with the rituals, the liturgy, his knowledge of the Bible and of the Talmud. In one respect he stands below many a family doctor who knows that he has no time to study and to read about the newest discoveries in medicine. The modern rabbi is not even aware that reading and studying modern philosophy and even science could be relevant to the performance of his duty. On the contrary, he may—just like Rabbi Heschel—look askance at all these novelties because he has been taught at the seminary to distrust them.

Frequently he is not aware that his audience may count many an intellectual who keeps abreast of the progress in science and who is keenly aware of the extreme discrepancy between his knowledge and the antiquated views of his rabbi. He rebels against the sentimentality of Rabbi Heschel's sermons, which could not be a substitute for true enlightenment. The modern slogan for an acceptable sermon should be credibility. Alas, much creed of many a religion is not synonymous with the credible.

True enlightenment and nothing else may save religion. This may sound like cheap propaganda. Let me remind you that the word "propaganda" came from the name of one of the Roman Catholic Church's central committees, De Propaganda Fide—For the Sake of Spreading the Faith. Spreading the faith does not need to be a dirty word as long as we do not write "faith" with a capital "F." One should not spread faith in a dogmatic doctrine but in enlightenment. Honest propaganda is, after all, teaching, and only reliable information is worth being taught. It is thus obvious that a rabbi should excel in intelligence, open-mindedness, and clarity of expression. This in addition to his moral integrity.

I believe that such an ideal teacher should reeducate his congregation in three stages with the aim to restore their interest in Jewish religion—not just in the feeling of Jewish solidarity—because it is religion especially and not national solidarity that suffered in the Western world since the Age of Enlightenment. These three stages are:

1. Teaching his flock the arguments that compel us to acknowledge that God exists;
2. teaching his flock the arguments that the world observed through our senses, the world of matter and space, is the product of our mind, and not the other way around—the matter of our brains cannot create mind; and
3. teaching that the reality of mind implies that the materialistic viewpoint of the nineteenth century has to be rejected and that this implies that our moral sense, which is an element of the mind, has to be taken very seriously.

As to the first stages in this teaching, the teacher—the rabbi—may cull his subjects from the first chapters of this book. His sermons may sound like lectures in astrophysics but the ultimate object, the conviction among his pupils that God created the world, is the principal condition for the revival of devotion that suffered immensely from Marx's and Darwin's philosophies.

The substance of his lectures is clear. He may start with the question, "If there has been a day in the infinite past?" His audience may first hesitate how to answer this question but the implication that an eternity would have to elapse after that day in the infinitely remote past before his listeners would be born, that, in other words, they would never have been born, must convince them that a day in the infinite past is absurd. Time must have a beginning, this world is unthinkable without time, hence the world has been created and creation is unthinkable without a Creator. The rabbi may underpin the new revelation with whatever his audience may understand about astrophysics.

The second stage in the rabbi's task is to overcome the superstition of our age that the processes in the matter of our brains create our mind. The first part of this book again provides the vital arguments. After explaining to his audience that things are, in reality, not as we see them, he may proceed to the subject of the immediacy of thoughts and feelings, and the indirectness of sense perceptions (either directly through the senses or through the interposition of instruments. Remember that thoughts and feelings are direct and not separated from the person by any intervening medium). The ideal rabbi has to explain that our senses serve us to guarantee our survival but that they are not good enough to convey to us, directly, the real character of the external world.

And what about biblical miracles? One of the most difficult tasks of the rabbi is how he should relate biblical miracles to the very few data of modern parapsychology that—whether explainable or not—have been verified as being absolutely reliable from a scientific point of view.

A miracle that has been explained or even brought within the realm of the explainable is not a miracle anymore and loses its effect as a miracle. Furthermore, we know, today, that the wildest apocalyptic hallucinations of the Revelation of St. John the Divine are child's play compared with real processes taking place in the astral world and that we are even able to perform them at (evil) will in atomic warfare. The miracles in the Bible do not impress us anymore, at least not the miracles that appeal to the craving for sensational spectacles. As to the other biblical miracles, I am personally convinced that they are inspired by extrasensory perceptions, such as premonitions and clairvoyance, phenomena that were certainly known of in ancient times.

The narrator made use of them as a means to underline important messages, often of supreme moral interest. Should we regard this form of accentuation of the biblical message as an essential element in Holy Writ, or should we disregard miracles altogether? I think that we had better leave the answer to the judgment of the reader, but some among the Bible-readers, especially ardent believers, tend to regard biblical miracles as supporting evidence of Divine truth.

J. Guttman quoted Moses Mendelssohn's rejection of this view in

Philosophies of Judaism: "The truth of any religion is found, not in any outward miracle, but in the inner truth of the doctrine," or "no miracle can attest to the truth of any faith that is unable to withstand the probings of reason," and "miracles cannot give birth to conviction." I go even further. I regard the message of the Bible credible, notwithstanding the biblical miracles.

And thus you may ask me my opinion. How should a rabbi handle biblical miracles in his sermons? I think he should render them exactly as they are narrated in the Bible. He should never mutilate Holy Writ but he should never adduce miracles as arguments to the veracity of Divine Truth. The description of miraculous events in the Bible are meant to draw attention to very important events, I am not obliged to adduce examples.

The second stage is indeed a rather complex structure, composed of convincing arguments that mind is reality, and matter a form in which mind is perceived through the senses, and that secondly, our mind and our senses are inadequate instruments to pass judgment on the veracity of biblical miracles. It is as if God raises His voice to stress His point and He uses the miraculous as his megaphone.

The third and upper layer of the foundation for the new religious philosophy, the analysis of our moral sense, covers the two lower layers—the knowledge that God created the world and the conviction that mind is reality and matter is mind observed through the spectacles of our senses. The rabbi has to presume that his pupil is amenable to moral motives. Moral sense cannot be taught, it must be hidden in the mind and the rabbi can only bring its essence to the surface of the consciousness by explaining that the extreme importance of moral virtue follows from the reality of God and the reality of mind that He created. The rabbi will soon be aware if there is something to uncover, if he does not have to deal with an incurable scoundrel who despises moral arguments. Whenever this may turn out to be the case—it is only a very small category—he has no choice but to ignore him.

The rabbi has to face other problems. He may hit against an artificial wall around the soul of an individual, against the induration of false ideologies. Atheists, for example—most are fakes—may feign that they are impervious to the most convincing arguments that the world has been created by a Creator. Then there are the deists who consult their intellect but never their heart. They may shrink from the consequences of stage three and they may refuse to face their inner conviction that moral virtue is something special. And then there are the materialists, the Marxists, and others who may abide by the superstition that thoughts are the secretion from the brains, something of no consequence and fanciful, and that therefore moral arguments, which are thoughts, are equally hallucinations and that their reality should be dismissed. I believe, however, that most of them follow just contemporary fads—such as existentialism—and that they may be convinced of their mistake.

A freethinker who has been convinced of the truth that God created the world, that mind is the reality behind matter, that moral virtue—God's gift—justifies, is not yet a devout believer; but he has come very close to it.

And why does the three-stage foundation of my religious philosophy fall short of religion as experienced by pious Jews? Because a really pious Jew would squint at me as if I were a building contractor who just completed the basement and has no money to erect the edifice. Let me quote once again Leo Baeck: "For Judaism, religion does not consist simply in the recognition of God's existence. We possess religion only when we know that our life is bound up with something eternal, when we feel that we are linked with God and that he is our God." I would have preferred the words "bound up with something timeless" as it sounds better than "bound up with something eternal." We associate the eternal with infinite and infinite time does not exist.

What Baeck meant is that we should not only know that God created the world but that we should also feel that He created us, and live accordingly. Indeed, I have built a foundation of three layers but I am not an architect. Various kinds of Judaism may be erected on top of it—Orthodox, Conservative, Reform, and so on—it is up to you and the rabbi. But there is one Judaism that my foundation cannot sustain, and that is Sherwin T. Wine's Humanistic Judaism. A Judaism without God is not even a religion, never mind a Judaism. Jewish atheism is exceedingly inconsistent. Judaism has to be defined as the belief in one God and the Mosaic laws. Wine's Humanistic Judaism cannot be sustained by the first layer of my philosophy that God created the world out of Nothing.

Materialism, Historical Materialism, and Marxism are at odds firstly with the second layer. For them matter is only reality and mind is nothing but an unessential secretion from the matter of the brain. They do not realize that matter cannot be perceived without the actions of our senses and our mind; that we do not have any reason to conclude that matter exists or anything would exist if we would have to do without our senses and without our mind. We would wind up with a philosophy of absolute nihilism.

But my three-layered religious philosophy would not sustain a Christian philosophy either. All the Christian churches, Catholic and Protestant, believe in justification by faith along with meritorious acts. This rebellion—it is against the third layer of my philosophy—is mortal sin.

I have pointed out why a man brought up in a Western civilization cannot embrace Islam. This leaves us to decide what form of Judaism we are ready to accept, and I admit that I am not sure how to navigate between the three denominations, though I am not an Orthodox since orthodoxy proclaimed the Talmud as Holy Writ. Furthermore, I shy away from committing myself to follow the rules of the Reform because of its very unpleasant history. And as to the denomination of the Conservatives, I feel that there is something wanting, something I looked for in vain in Reform Judaism as well. It is the absence of a sound religious philosophy, the odd feeling that they float in the air, that they are modifications of the old Orthodox, adjustments of rituals and liturgy to changed circumstances. They are not built on a systematic religious philosophy.

I referred in parts to a remarkable booklet, *The Search for Jewish Theology*, by Bernard Jacob Bamberger. His search included a quest for a solid religious philosophy on which he could base a Jewish theology. The result is not as satisfying as one should wish. He based his knowledge of God on the transcendent source of moral strength. We should refer this truth to the third layer of a rabbi's education, but Bamberger ignored the first layer. I gather that he was ignorant of the valid Cosmological Argument and of its scientific corroboration. His work is, notwithstanding this defect, a noteworthy attempt to erect the principles of Jewish theology on a firm and rational religious philosophy.

While Bamberger based his knowledge of God principally on the intuition that moral virtue is a divine gift, we have seen that the rabbi has to convince his people of what Bamberger called "the transcendent source of moral strength" and we called this the third and upper stage of reeducation. We start with a much deeper level, the first and lower layer of reeducation, which is the knowledge of God based on the Cosmological Argument and its corroboration. In and between the lower and upper stage, or layer (convincing the people of the supreme importance of mind as the reality behind the external world of matter), may be regarded as the natural and rational deduction from the essence of the mind-body problem.

The deduction that matter is mind seen through the senses has nothing to do with the deduction in the lower stage that God created the world. But a most essential conclusion may be drawn from the two deductions together, that God created a world of mind, a spiritual world. It is furthermore obvious that the conclusion of the third stage, the absolute sovereignty of moral sense, does not follow at all from the conclusion drawn in the second stage. It is true, however, that the conclusion drawn in the second stage, that "mind does matter because mind creates matter" immensely strengthens our confidence in the supreme value of moral virtue.

And now we bind the three layers into one. We related the lower layer to the upper: God created the world out of Nothing and He created everything, moral virtue included, which we experience as the acme of Creation. And this explains why we feel moral virtue as a divine gift.

What is expected from this three-stage reeducation? The triple-layered foundation of a new philosophy that is supposed to restore the confidence of the rabbi's flock in the meaning of Judaism? Indeed, what do I mean with confidence in the meaning of Judaism? Certainly not loyalty to the Jewish people; a Jew who shows that he is not loyal to the Jewish people is a traitor. What I mean is loyalty to Judaism, and we have seen that Jews are not unanimous as to their obligations how to observe the commandments. There are even Jews who do not observe any commandment, they just behave in accordance with their innate intuition about good and evil. The rabbi confronts a wide scope of variant points of view. If he may ask you to be loyal to Judaism, you would immediately challenge him to tell you which among the various denominations of Judaism he has in mind. His choice defines the apparently slight, but in fact important, differences in ritual, liturgy, worship and service.

Let us in this concluding chapter return for the last time to square one of the opening chapters. To live a temporary life in a world of universal time and sensory space is the experience of everyone. That this world began as a dimensionless Singularity is the knowledge of the enlightened. The Singularity was not somewhere because there was no space. The Singularity had no being in duration because there was no time. The Singularity did not consist of energy because energy had not yet been created. The Singularity was Nothing and from this Nothing the world was created by the Nothing we call "God."

The permission to worship and to serve the Nothing is a concession to human clumsy helplessness, but this permission is granted under the condition that we realize what "worship" and "service" mean.

Yearning for God is not the piety of the devout but the sinful impudence of the mystic. The truly devout worships and serves God by his gratitude. The intensity of his gratitude expresses the strength of his piety, which is not bounded by limits; but the form through which he expresses his gratitude to God is limited by his whereabouts at a special moment, defining him in space and time. The worshiper is supposed to worship the timeless and spaceless at a certain moment, and on a certain place. It is a bewildering situation he tries to solve in the following way.

The moment is the moment of prayer and the place is the house of worship, or any other place of worship of his choice. But before we take the plunge into the depths of the world of observant religious believers, a few words about the others.

I am aware that quite a few among my readers feel a little inclination to look beyond my cogent argument that God exists and has created the world. Some among those who stand aloof had first to overcome a certain reluctance before they were ready to accept this obvious truth that all the clear-minded would readily admit. They do not believe, however, that this conclusion would entail any further implications for themselves and they are convinced that these discoveries do not touch their conception of life that challenges the meaning of the world and of their own existence.

We have encountered them before. They are the deists. They live in a kind of mental dualism. The existence of God is one fact and the existence of moral values another, but there is not interconnection between these two facts. On strictly scientific grounds their concept is unassailable. They would certainly not call it a religion.

Deists have their eyes glued on the heartless descriptions of the world of sense experience and they shun every thought about the value of values, especially when they face the world of natural science. I know that many are aware that their own spiritual would, which they see by introspection through the eyes of their mind, is strangely at odds with the world of the physicists. But they rarely try to collate these two worlds and they are at a loss how to harmonize them into one consistent and satisfying system, because they fail to realize that the world as one sees it through the senses is not the world as it is.

We have met even worse cases: those who dismiss moral values altogether, those who regard moral values as nothing but "bourgeois prejudices." They got this wisdom from Karl Marx. Do they really believe he behaved consistently according to his "faith"; that he—in all sincerity—had a deep contempt for those who believe in the essence of moral values?

If so, why did he dedicate his life for what was obviously a moral cause? What are these three thick volumes of *Das Kapital* else but an effort to restore the dignity of the workman? And is not this effort a response to Karl Marx's innate moral obligation? What was actually his message? "You rich of the world. Stop raving about moral values. Look what you have done. You have debased your comrades who produce the needs of this world. You were supposed to guide them, but you exploited them. Try first to redeem them from their present state, and let us then talk about moral values, all in unison."

But Marx did not have the courage to formulate his message in these words. He was so blinded by his hatred of the rich that he could not look into his own heart, and the Marxists blindly followed a blind man.

Classical Marxism is a perfect example of a movement that professes the nonexistence of morality and nevertheless fights for a moral cause out of deep-seated moral conviction. But we have seen that it is easy to fight classical Marxism. What to do, however, with the many who admit that moral agreements play a tremendous role in our behavior, but maintain that moral sense has nothing to do with God? We have seen that deists cannot be fought with the arms of pure reason because mortal sense cannot be explained with the arguments of pure reason.

In this single combat the deists have already chosen their arms: pure reason and nothing else. And therefore I believe that it is of little use that theists would try to convince deists, because if somebody chooses to be a deist out of a defect in his sense of morals, he can never be convinced. We have to forego further discussion and admit among ourselves that a deist, who knows that God exists, will never worship or serve Him. This bitter disappointment did not bring me in a very complacent mood. We have declined the dual because we both agree that we live in two different worlds.

Do not believe that I may proceed with ease after I have confronted the great majority who willingly followed me, at least to the point that moral virtue cannot be dissociated from God. A great many deists draw the obvious consequences in their search for a contact with their Creator. We have seen that three points of view prevail: (a) the world is God—the imminent view; (b) the world is part of God—the eminent view; and (c) the world has been created by God—the transcendent view.

You will remember that we dismissed the imminent view out of hand. The world has been created out of Nothing and if the world should be identified with God, God would have been created out of Nothing.

It is harder to dismiss the eminent view that claims that part of the Nothing turned into a something which is the world. If I am not hard of hearing, it would mean that we are supposed to be the scintillating sparks of God's Own spirit. My objection is that even the eminent view regards God as the perfect Nothingness devoid of attributes and it is hard to see how a world of attributes may be "part of" God. Furthermore, if we would be His sparks our only imperfection would be our finite state in time and space. In short, there are two possibilities. Either we bear God's perfection, which is not an attribute, and this is an absurdity, or we are imperfect—which is obviously the case—and therefore not even sparks of God. Seen from a more practical view: how may we expect that the sparks of God Himself compete one with the other, even fight one against the other? I believe, therefore, in a much more modest role of mankind...we are God's creation and not His tiny parts. This brings us, alas, back to the edge of the abyss of transcendence that menaces to separate man irrevocably from his Creator. Let us this time be on our guard not to be toppled down by the deists.

Let us first of all admit that any retreat from the brink of deist transcendence is a concession to "emanism." But this step is not necessarily a concession to mysticism as long as we keep within the well-defined borders of reason. If we are obliged to introduce an element of emanism into the view of absolute transcendence—if only to secure ourselves against the maddening loneliness and helplessness in our human situation—it must be a tangible element.

We have seen that the only element that may receive consideration is moral virtue. Moral virtue is a tangible asset though not an element that can be explained by reason. We recognized that all the other intermediaries between God and man proposed by other religions, whether Christ or Mohammed, are mere phantasms.

After having accepted that this view is the correct one, we have still to cope with four questions:

1. If we received our moral sense directly from God, does this mean that we must render obedience to God?
2. If so, what does God command?
3. How are we expected to express our obedience?
4. What has the liturgical form, as prescribed by the various religions, to do with the expression of our obedience?

The answers to the first two questions are already implied in what I have explained more than once. We render obedience to God if we heed the moral rules as circumscribed in Heymans's principle of objectivity which defines our attitude and behavior towards our fellow men. This is the answer to the first question and it implies the answer to the second question. We are commanded to make use of our moral sense. But many Jews may try to point out that God commands a great deal more, and they mean that we are commanded to perform acts that are not related to our fellow men but to God. They would, for example, point to Exodus with all the details of how we are commanded to build the Tabernacle. Let me put them on the right scent. These obligations are meant to instill due respect for the *pièce de résistance* in the Tabernacle, the Decalogue hewed by God's hand in the two tablets of stone that contain the moral Message, identical with the moral Message that He hewed into the heart of mankind and which He made readable to him in the garden of Eden. And this may teach all of us that all these commandments of Exodus 25, 26, and 27, which are seemingly meant to serve God alone, are nevertheless meant to remind us day and night of our moral duties.

This answer would not answer the question why all these prescribed paraphernalia that are commanded to encompass the two holy tables as what looks like nonessential adornments? I can make their function understandable after a short introduction on values.

We talk about cultural, artistic, national, and even financial values. It would be a gross understatement if we would say that all these so-called values do not come up to moral values. It would be more to the point if we would claim that compared with moral values, all the other values are of a world apart. They have with moral value only the name value in common.

I admit that things of cultural, artistic and aesthetic value embellish the world but I feel also that a world without moral values, without moral standards, would be worthless. "In terms of human values, the world of creation is irrelevant and meaningless," said Professor Yeshayahu Leibovitz on Israeli Television, 6 March 1987. He pointed at the two tables of stone that represent the Torah. Let me add that an irrelevant and meaningless creation—the principal article of faith among the deists—would point to an irrelevant and meaningless Creator. Is there a decent man ready to endorse this blasphemy?

Let us return to the term "value" in general. On the point of values man is totally corrupt. Let me give a rather innocent example. We buy objects of art. The artist parts with his masterpiece and the purchaser parts with some of his riches. It is generally felt that the artist makes a greater sacrifice than the new owner, and we fail to realize that our intuitive distinction means that we know in our heart that the deal was not quite honest. A real artist is not reluctant because of the low offer, but because the artist and the buyer pay in different currencies. The artist pays with his heart and the purchaser with his pocket. That is why it is so difficult to clinch the deal. There is no exchange rate between financial and artistic values.

We can now take a much more painful example: the reparations by the Germans to the Jews after World War II, or any other court-ordered reparation or payment. What is the financial value of a human life? The art dealer could at least claim that he secures the livelihood of the artist. Indeed, somebody must be responsible for the sustenance of the artist, but this consideration does not provide for an exchange rate in values.

What about moral values? I believe that moral value is only payable in divine currency. A religious Jew wants to express his gratitude for the opportunity to fulfill his moral duties as God has commanded. To show gratitude to God means to pay tribute to Him. This payment can only be made in the currency of divine moral value, in the coins of prayer, of blessing and thanksgiving.

What Bernard J. Bamberger had to tell us on this subject in *The Search for Jewish Theology* is a simplified formulation of the obvious. When the devout loses himself in prayer, he turns—mostly unwittingly—his monotheistic belief into a belief in the "unipersonal." Instead of professing the one God, he addresses his Creator as a Person, because he cannot shake hands with Nothing. We have seen that this universal behavior of the devout is a symptom of an excusable human weakness. The other human shortcoming is that he does not leave it at mere thanksgiving but that he adds requests for a favor bestowed by God.

Bamberger commented on petitionary prayer that a supplication for something entails moral and religious difficulties. He adduced the well-known example of two "warring nations each praying to the same God for support in the project of defeating and destroying each other."

Judaism knows of three kinds of prayer. Prayers commanded, prayers allowed, and prayers forbidden. I have my own personal views on the forbidden prayers. I believe that a great many customary supplicatory prayers should be prohibited. In fact, I would allow a petition for strength to fulfill our moral duties. What do we observe in practice? That the genius who can muster sufficient fantasy and inspiration to compose his own prayers is an exception. But a man is not judged by his moral virtues. Usually we see how the simple pious browses through the Book of Psalms until he hits upon a text of a supplication that he believes will do the trick to mollify the Supreme. There is hardly a habit so demoralizing as falling on your knees whenever a misfortune, a small stumbling-stone, distresses you. It saps your moral strength.

Let us now turn to the second parameter, the "place" of worship, which also serves to pursue the point where the following question had been asked. "Why all these prescribed paraphernalia, which are commanded to encompass the two holy tables, seem to be nonessential adornments?" Let us go back to Exodus after Moses came down from Mount Sinai.

All the Jews had were the two tables of testimony in Moses' hand. But what about their notion of worshipping? They were not yet in possession of the Torah, or a Talmud. What they had witnessed in Egypt of "worshipping" was not worth remembering. But they remembered it, and how much they had hated their oppressor, the majesty of all that gold and the colors of the murals in the Egyptian temples was still fresh in their minds. How could one ever expect that the Chosen would be ready to serve God without such pleasant surroundings?

God had to reckon with this pagan infection of their minds. But slaves, even liberated slaves, do not possess gold or silver, nor brass, and God had to provide the Children of Israel with these needs through a bit of cunning, a bit of mischief. "Ye shall spoil Egypt," He ordered, "by borrowing from you neighbor jewels of silver and jewels of gold and raiment (Exodus 3:21–23)." These were borrowed valuables that the Jews neither intended nor could return. I believe that these riches were meant as a form of reparation for the suffering in Egypt. I admit that God's ways are inscrutable; it may even be seen as a piece of His providence. His people were amply supplied the materials to decorate the two tables of stone to their heart's content.

But we still have a serious problem. God did not grant His people a free hand to build His sanctuary, the Tabernacle, to its own taste. Instead of His high consent we find three chapters full of directive orders how to go about it. God is the architect, Moses the contractor, and the people of Israel the masons. Why?

It all boils down to the same point. They were worse off than the Negro slaves after the civil war in America. They were still near-pagans. If this Chosen People believed in dignified surroundings to serve

God, if it even believed that it could not serve him without them, God had to guide the construction work down to the smallest detail. To adorn the two tables of stone is not the same as decorating a birthday cake. It is to embellish the highest value of values. The highest value of values is its own embellishment. Every man-made addition may violate it. Every excessive piece of beauty may divert the attention from the central object—the Ten Commandments.

Too much gold in the hands of the most sincere and well-meaning people in need of a religious outlet, but lacking proper direction, tempts to idolatry. It builds a golden calf behind the prophet's back; but let us regard this just as a sad interval.

There is an enormous difference between the building of the Tabernacle as ordained by God, explained in Exodus 25 and onward, and the free enterprise of the construction in Jerusalem of King Solomon's Temple. As it is said: "And it was in the heart of David my father to build an house for the name of the Lord God of Israel."[5]

It became a construction to the taste of His Majesty the King. Exodus 25, 26, and 27 are prescriptions how to build the Tabernacle; I Kings 6 is a description of King Solomon's performance. I could not find any indication that he had been guided from above. He carried out the will of his father. But God did not test Solomon's good taste, He judged him by his moral sincerity. The King did not even need to open his mouth to hold his oration at the inauguration. God knew what was in his mind: "—and it came to pass, when the priests were come out of the holy place that the cloud filled the house of the Lord, so that the priests could not stand to minister because of the cloud: for the glory of the Lord had filled the house of the Lord."[6]

Was it a real cloud or was everyone moved to tears when "the priests brought in the Ark of the Covenant of the Lord unto his place—to the Holy of Holies, under the wings of the cherubims"? Was the glory of the Lord outside or in their hearts?

We may imagine that even the king stood there for a moment struck dumb and overcome by the solemnity of the moment. He opened his inauguration address with laudatory thanksgivings to the Lord Who brought the people of Israel to the Promised Land. But all of a sudden there must have been a catch in his voice and he introduced a sentence that was much too beautiful to have been prepared in advance. It was a spontaneous stammering: "But will God indeed dwell on the earth? Behold, the heaven and the heaven of heavens cannot contain thee; how much less this house that I have built?"

For the Nothingness Who created space and time and being outside
time and space there could be no place in a house built by human hands.
In mid-speech, King Solomon must have had the feeling that all this
transitory splendor of his making, this house of threescore cubits length
and twenty cubits breadth overlaid within with pure fine gold, even the
cherubim within the "*dwir*," sank into His mighty Nothingness. His
royal majesty had made a mistake. He had confounded values. He had
tried to pay tribute in gratitude for the divine gift of moral values in the
currency of gold.

Why did God honor this check? Because of the sincerity of the King's
piety. Values are of a tricky value. We came to realize this long after
King Solomon. The truth is so confusing. There are so many kinds of
values. What about the value of life, for example? Is the value of life
not closely related to moral values?

Furthermore we often confound two different kinds of values: the
values of human talent and faculty are not the same kind of "values" as
the value of their products. There is a fundamental difference between
the value of artistic talent and the value of pieces of art, between the
value of intellectual faculties and the value of gathered knowledge,
between the value of moral virtue and the acts of moral behavior.

We began to understand this after a long period of maturing insight.
On our way we met with many stumbling blocks. We have seen, for
example, that Spinoza believed in the superior value of human intellect; and not even that. He believed that the acme of value was the
value of gathered knowledge, knowledge about God. This was a fatal
mistake. We will not shed a tear for someone who died on account of
his high intellect. Hitler was surrounded by many highly intelligent
advisers.

What may have caused Baruch's mistake? I believe because of the
special character of "knowledge," it is the only value that a professor, a
teacher, may bestow upon others without parting with it. Digesting new
knowledge, writing about it, teaching it, does not make one poorer but
richer. But even the greatest saint cannot bestow his moral virtue upon
others. That is the great misfortune of mankind.

We also have to face the fact that moral virtue is a rather ineffective
asset whenever it is not assisted by intellect. There are quite a few
ignorant saints who do much harm, notwithstanding their best intentions. It is only in the rare and precious cases where the highest moral
virtues merge with a sharp intellect that man raises himself above his
fellow men. These people are called "men of wisdom" and the seam

between their ego and superego is marked by the high value of their tact. We are told that King Solomon was such a man.

What happened thereafter is a matter of interpretation. We may surmise that the daily service in the Temple soon became a rut after it had been learned by rote. The worshipers could not help to be distracted from the humdrum by all the gold. What could be the meaning of all these glittering sculptures? Did these twelve oxen, carrying "the molten sea, ten cubits of the one brim to the other" symbolize twelve adult golden calves? It became too obvious that the Temple of Solomon was little better than a pagan adornment around the great divine Message written on the two tables of stone concealed in the Ark.

The Temple was overshooting its mark. God appeared a second time in a dream, and He admonished the king that this may happen: "Keep My commandments and my statutes which I have set before you" and be on guard that they do not "serve other gods, and worship them."[7]

Feasting your eyes on gold, silver, and brass leads to idolatry. The king brought the calamity upon himself and his people by the ostentatious display of his poor taste. God did not order him to carry out his duty, the will of his father, by erecting a pompous building of worldly splendor.

It took more than two centuries before the waves of a great empire lapped at the shores of the Judean kingdom and swept this enormous Solomonian sandcake on Mount Moriah away. But Nebuchadnezzar and all the other following despoilers could no longer harm the Children of Israel. The records in the Bible are on this point imprecise. What exactly was Nebuchadnezzar's offense? In II Kings 24:13 we read: "And cut in pieces all the vessels of gold which Solomon the king of Israel had made in the Temple of the Lord, as the Lord had said." But in II Chronicles 36:7 we find: "Nebuchadnezzar also carried of the vessels of the house of the Lord to Babylon, and put them in his temple at Babylon." And this record, which is not the same as the alleged "cutting in pieces" is corroborated in Ezra 1:7, which tells about the first steps taken to build the Second Temple.

But we may regard this a minor detail. A much more serious problem has been left open. What happened to the ark and the two tables of stone? My efforts to find any record about them and their fate in the Bible were unsuccessful. They may have been lost. Did this matter very much? I do not believe so. Meanwhile God performed another miracle: Max I. Dimont pointed in his *Jews, God and History* to the theory of Oswald Spengler that a nation in exile must perish; it is a "law of nature." Why do Jews defy this law? Why did they survive—as Jews—in Babylonia? They survived because they canonized the Torah when they were in exile.

This momentous act turned the two tables of stone with the Holy Message into a holy scroll, written in a holy language and containing all the moral obligations that live in the heart of the virtuous. Even the destruction of the Second Temple by the Romans could not harm them any more, not even the Christians who tried to corrupt the Jewish mind by persuading them to dispense with morals. Look what St. Paul dared to write to the Corinthians about the veil that covered Moses' radiating countenance when he descended with the two tables from Mount Sinai: "We use great plainness of speech; and not as Moses who put a veil over his face, that the children of Israel could not steadfastly look to the end of that which is abolished."[8]

And what is "the end of that which is abolished"? We find the answer in Romans 10:4: "For Christ is the end of the law for righteousness to every one that believeth." And the conclusion is: "Nevertheless when it shall turn to the Lord, the veil shall be taken away.[9]

I call this the double rape of Moses by Jesus Christ. First Christ pulls down the veil from Moses' face and then he forces the two tables out of his hands. The veil was there to protect the bitten conscience of the Jews after the worship of the golden calf from the divine radiation shining from Moses' face and "the end of that which is abolished" is the abolishment of moral ends and the abolishment of jurisdiction by moral conduct.

This is worse than merely an attack on Judaism. It is an attack on God, donor of moral virtue, the highest value of values. God did not weep went the Temple when up in flames: "Will I eat the flesh of bulls and drink the blood of goats?" as it is written in Psalm 50:13, or "To what purpose is the multitude of your sacrifices unto me?" as written in Isaiah 1:11. The Temple is gone and the two tables may have been ground to dust in the process. But meanwhile the holy text has been copied repeatedly into thousands of scrolls and millions of printed Bibles. All is written on most worldly parchment, printed on plain paper, and—just to keep up with modern science—saved on floppy and compact discs. Holy Writ is Holy Writ, just as the two tables of testimony were earthly stones. They did not descend in God's hands from heaven like the first ones which Moses "cast out of his hands, and broke them beneath the mount" and which were the "work of God." The tables that finally found their way into the Ark were hewn by God's mortal servant, Moses, by his two human hands, it was only God who engraved the words. God had said to Moses: "Hew thee (two) tables of stone like unto the first and I will write upon them."[10] In modern language, "give me something to write on, just two stones." May this serve as a reminder, after the sad interlude of the "golden calf," that moral values are not a man-made idol.

The First and the Second Temples have been razed to the ground. We are told that the Third Temple will be the work of God Himself. Who knows what that may mean? Will we be able to see the Third Temple with our mortal eyes? Who needs a temple after all? The Ark and the two tables are not there anymore, but let us imagine that the engravings by God's Holy Hands are still present, written in the thin air above Mount Moriah in Jerusalem.

Notes

1. This definition is from the *Concise Oxford Dictionary*.
2. *Out of My Later Years*.
3. Free rendering from the Hebrew text by the author.
4. *The Perennial Philosophy*.
5. I Kings 8:17.
6. I Kings 8:10-11.
7. I Kings 9:6.
8. II Corinthians 3:12-13.
9. II Corinthians 3:16. This means "the heart" as it is said in II Corinthians 3:15 "the veil is upon their heart and the veil is done away in Christ."
10. Exodus 34:1.

Index